AF334038

Quantum Information IV

Also published by World Scientific

Quantum Information
Proceedings of the First International Conference
eds. T. Hida and K. Saitô
ISBN 981-02-3934-3

Quantum Information II
Proceedings of the Second International Conference
eds. T. Hida and K. Saitô
ISBN 981-02-4317-0

Quantum Information III
Proceedings of the Third International Conference
eds. T. Hida and K. Saitô
ISBN 981-02-4527-0

Proceedings of the Fourth International Conference

Quantum Information IV

Meijo University, Japan　　　　27 February–1 March 2001

Edited by

T. Hida & K. Saitô
Meijo University
Japan

World Scientific
New Jersey • London • Singapore • Hong Kong

Published by

World Scientific Publishing Co. Pte. Ltd.

P O Box 128, Farrer Road, Singapore 912805

USA office: Suite 1B, 1060 Main Street, River Edge, NJ 07661

UK office: 57 Shelton Street, Covent Garden, London WC2H 9HE

British Library Cataloguing-in-Publication Data
A catalogue record for this book is available from the British Library.

QUANTUM INFORMATION IV
Proceedings of the Fourth International Conference

ISBN 981-238-020-5

Printed in Singapore by Uto-Print

PREFACE

During the past four years we have had the most fruitful time for the research aiming at the development of quantum information theory. Here, in this volume, are reports on some of our main results that have been obtained so far by the members of the Academic Frontier Project at Meijo University. The papers in this volume are based on the lectures presented at the Fourth International Workshop on Quantum Information at Meijo University for the period February 27–March 1, 2001. The editors are pleased to accept all papers for publication in this volume at the suggestion of the referees.

It is noted that there has been a great achievement of the theory in both white noise analysis and quantum probability theory starting from the seventies of the last century. It was our original intention to develop these two theories and further to establish unification of white noise analysis and quantum probability in order to explore new important directions including quantum computation and quantum information theory. We are very happy to see that this idea has been realized to a certain extent. In addition, a new impetus has come to probability theory from its strong interaction with quantum dynamics. Thus, we are encouraged to launch a new direction of our research.

The Fourth Workshop was really a good opportunity to report the recent results on various hot topics and exchange new information in the fields of interest. At the same time we have been trying to proceed the revolution of the established areas. In this sense the workshop was quite exciting and has received good reactions from all the participants.

As for the white noise theory, which is basically connected with the quantum information theory, is now about to meet its renaissance. Having good collaboration with quantum probability, as well as more intimate relationship with physics, we are sure that we are steadily approaching our goal. The quantum probability has now built a bridge between its results and quantum computer. We are most hopeful of success.

Our Academic Frontier Project will be closed at the end of the 2001 academic year. The series of the publication "Quantum Information" shall also be closed. We promise to edit the last volume with a new idea so that the quantum information theory will develop towards new directions and the theory becomes more popular in science and technology.

The organizers are grateful to the speakers at the workshop who contributed papers for this volume.

December 30, 2001
Takeyuki Hida
Kimiaki Saitô
Meijo University

Quantum Information IV (pp. 1–37)
Eds. T. Hida and K. Saitô
© 2002 World Scientific Publishing Co.

A QUARTER CENTURY OF WHITE NOISE THEORY

HUI-HSIUNG KUO

*Department of Mathematics, Louisiana State University,
Baton Rouge, LA 70803, USA*

ABSTRACT. T. Hida created the mathematical theory of white
noise in his Carleton Mathematical Lecture Notes "*Analysis of
Brownian Functionals*" (vol. 13, 1975). This theory has since
been extensively developed and is now recognized as an important
branch of stochastic analysis in the MSC 2000 subject classifica-
tions as "white noise theory" with the code number 60H40. We
will review the shaping of white noise theory and the progress in
various applications during the last quarter century.

1. Pre-theory period (white noise without a theory)

White noise is a sound with equal intensity at all frequencies within
a broad band. The sound in a large orchestra before the conductor
raises his baton is an example of white noise. Mathematically it is
informally defined as a stochastic process $z(t)$ such that it is independent at different times and has identical distribution with mean 0 and
infinite variance in the following sense:

$$E(z(t)z(s)) = \frac{1}{2\pi} \int_{\mathbb{R}} e^{i(t-s)x} \cdot 1 \, dx = \delta_0(t-s),$$

where δ_0 is the Dirac function at 0. With this definition of white noise
$z(t)$ we can informally derive the following equality for a real-valued
function $f \in L^2(a, b)$,

$$E\left(\int_a^b f(t)z(t) \, dt \right)^2 = \int_a^b f(t)^2 \, dt.$$

But what is the definition of the integral $\int_a^b f(t)z(t) \, dt$? Moreover, is it
possible to define $z(t)$ for each t?

We can think of $z(t)$ as the derivative $z(t) = \dot{B}(t)$ of a Brownian motion $B(t)$. Then by regarding $z(t) \, dt = dB(t)$, the integral $\int_a^b f(t)z(t) \, dt$
can be defined as the Wiener integral $\int_a^b f(t) \, dB(t)$. However, since every Brownian path is nowhere differentiable, $\dot{B}(t)$ does not exist for
any t.

More importantly, consider a stochastic process $f(t)$, is it possible
to define the integral $\int_a^b f(t) \, dB(t)$ or even the integral $\int_a^b f(t)\dot{B}(t) \, dt$?

It is well-known that the Itô integral $\int_a^b f(t)\,dB(t)$ is defined for all nonanticipating stochastic processes $f(t)$ with almost all sample paths in $L^2(a,b)$. But in order to define the white noise integral $\int_a^b f(t)\dot{B}(t)\,dt$ we really need to give a rigorous definition of $\dot{B}(t)$ at least for almost all t.

2. Creation of white noise theory (1975)

The mathematical theory of white noise was created by T. Hida in his 1975 Carleton Mathematical Lecture Notes [123]. In fact, five years earlier he already envisioned the white noise theory in his Princeton University Press book [116] by discussing Gaussian white noise, Poisson white noise, infinite dimensional rotation groups, etc. But it was in [123] that Hida proposed to use the family $\{\dot{B}(t);\, t \in T\}$ as a coordinate system to analyze white noise functions. Recently he has advocated calling $\{\dot{B}(t);\, t \in T\}$ a system of *idealized elemental random variables* as a result of reductionism.

To explain the original idea of Hida in [123], take the Schwartz space $\mathcal{S}(\mathbb{R})$ of rapidly decreasing real-valued functions on $\mathbb{R}$. Let $\mathcal{S}'(\mathbb{R})$ be the dual space of $\mathcal{S}(\mathbb{R})$. Since $\mathcal{S}(\mathbb{R}) \subset L^2(\mathbb{R})$ and $L^2(\mathbb{R})$ can be identified with its dual space by the Riesz representation theorem, we get a Gel'fand triple

$$\mathcal{S}(\mathbb{R}) \hookrightarrow L^2(\mathbb{R}) \hookrightarrow \mathcal{S}'(\mathbb{R}).$$

By the Minlos theorem there exists a unique measure μ on $\mathcal{S}'(\mathbb{R})$ such that

$$\int_{\mathcal{S}'(\mathbb{R})} e^{i\langle x,\xi\rangle}\,d\mu(x) = e^{-\frac{1}{2}|\xi|_0^2}, \qquad \xi \in \mathcal{S}(\mathbb{R}),$$

where $|\cdot|_0$ is the $L^2(\mathbb{R})$-norm. The elements in $\mathcal{S}'(\mathbb{R})$ can be regarded as $\dot{B}$ and for this reason the probability space $(\mathcal{S}'(\mathbb{R}), \mu)$ is called the *white noise space*.

For simplicity, let (L^2) denote the complex Hilbert space $L^2(\mathcal{S}'(\mathbb{R}), \mu)$. By the Wiener-Itô theorem each $\varphi \in$ has a unique decomposition

$$\varphi = \sum_{n=0}^{\infty} I_n(f_n), \qquad f_n \in L^2_{\text{symm}}(\mathbb{R}^n),$$

and the (L^2)-norm of φ is given by

$$\|\varphi\|_0 = \left(\sum_{n=0}^{\infty} n!|f_n|_0^2 \right)^{1/2},$$

where I_n is the multiple Wiener integral of order n.

Define a space $(L^2)^+$ of test functions and the corresponding space $(L^2)^-$ of generalized functions by

$$(L^2)^+ = \left\{ \sum_{n=0}^{\infty} I_n(f_n); \; f_n \in H_{\text{symm}}^{\frac{n+1}{2}}(\mathbb{R}^n) \text{ and } \sum_{n=0}^{\infty} n! |f_n|^2_{H^{\frac{n+1}{2}}(\mathbb{R}^n)} < \infty \right\},$$

$$(L^2)^- = \left\{ \sum_{n=0}^{\infty} I_n(F_n); \; F_n \in H_{\text{symm}}^{-\frac{n+1}{2}}(\mathbb{R}^n) \text{ and } \sum_{n=0}^{\infty} n! |F_n|^2_{H^{-\frac{n+1}{2}}(\mathbb{R}^n)} < \infty \right\},$$

where H^s denotes the Sobolev space of order s. Then we have the triple

$$(L^2)^+ \hookrightarrow (L^2) \hookrightarrow (L^2)^-.$$

The space $(L^2)^-$ contains elements such as the white noise $\dot{B}(t)$ as well as the renormalizations $:\dot{B}(t)^n:$ and $:e^{\dot{B}(t)}:$ for each fixed $t \in \mathbb{R}$.

The renormalization of $\dot{B}(t)^n$ is similar to Itô's idea of multiple Wiener integral, but the limit is taken in the generalized sense. Consider the case $n = 2$. Itô's idea is the following limit:

$$
\begin{aligned}
:B(t)^2: &\equiv \int_0^t \int_0^t 1 \, dB(u) dB(v) \\
&= \lim_{n \to \infty} \sum_{i \neq j} \big(B(t_i) - B(t_{i-1}) \big) \big(B(t_j) - B(t_{j-1}) \big) \\
&= B(t)^2 - t, \quad \text{limit in } (L^2).
\end{aligned}
$$

On the other hand, here is Hida's idea to define the renormalization of $\dot{B}(t)^2$ as a generalized function in the space $(L^2)^-$:

$$
\begin{aligned}
:\left(\frac{B(t+\Delta) - B(t)}{\Delta} \right)^2: &= \frac{1}{\Delta^2} \int_t^{t+\Delta} \int_t^{t+\Delta} 1 \, dB(u) dB(v) \\
&= \frac{1}{\Delta^2} \lim_{n \to \infty} \sum_{i \neq j} \big(B(t_i) - B(t_{i-1}) \big) \big(B(t_j) - B(t_{j-1}) \big) \\
&= \frac{1}{\Delta^2} \left[\big(B(t+\Delta) - B(t) \big)^2 - \Delta \right].
\end{aligned} \tag{2.1}
$$

The limit of Equation (2.1), as $\Delta \to 0$, does not exist in (L^2). But if we take the limit in the space $(L^2)^-$ of generalized functions, then the limit, denoted by $:\dot{B}(t)^2:$, exists and defines a generalized function in $(L^2)^-$.

3. Kubo-Takenaka's construction (1980)

In 1980 Kubo and Takenaka [260] constructed test and generalized functions on a general space. Let $\mathcal{E} \hookrightarrow E \hookrightarrow \mathcal{E}'$ be a Gel'fand triple.

Here E is a real Hilbert space with norm $|\cdot|$ and $\mathcal{E}$ is a nuclear space with norms $\{|\cdot|_p; \ p \geq 1\}$ satisfying the conditions:

(a) There exists $0 < \rho < 1$ such that $|\cdot| \leq \rho |\cdot|_1 \leq \rho^2 |\cdot|_2 \leq \cdots \leq \rho^p |\cdot|_p \leq \cdots$ for all $p \geq 1$.

(b) For any $p \geq 1$, there exists $q \geq p$ such that the inclusion map $i_{q,p} : \mathcal{E}_q \hookrightarrow \mathcal{E}_p$ is a Hilbert-Schmidt operator, where $\mathcal{E}_p$ is the completion of $\mathcal{E}$ with respect to the norm $|\cdot|_p$.

Let μ be the probability measure on $\mathcal{E}'$ with characteristic function

$$\int_{\mathcal{E}'} e^{\langle x, \xi \rangle} \, d\mu(x) = e^{-\frac{1}{2}|\xi|^2}, \quad \xi \in \mathcal{E}.$$

The probability space $(\mathcal{E}', \mu)$ is an *abstract white noise space*. Each $\varphi \in (L^2) \equiv L^2(\mathcal{E}', \mu)$ can be uniquely represented by

$$\varphi(x) = \sum_{n=0}^{\infty} \langle :x^{\otimes n}:, f_n \rangle, \quad f_n \in E^{\widehat{\otimes} n}, \tag{3.1}$$

where $:x^{\otimes n}:$ is a Wick tensor (see page 33 in the book [290]. Moreover, the (L^2)-norm $\|\varphi\|$ of φ is given by

$$\|\varphi\| = \left(\sum_{n=0}^{\infty} n! |f_n|^2 \right)^{1/2},$$

where $|\cdot|$ is the norm on $E^{\widehat{\otimes} n}$ induced by the norm on E.

For φ represented by Equation (3.1) and $p \geq 1$, define

$$\|\varphi\|_p = \left(\sum_{n=0}^{\infty} n! |f_n|_p^2 \right)^{1/2},$$

where $|\cdot|_p$ is the norm on $\mathcal{E}_p^{\widehat{\otimes} n}$ induced by the norm on $\mathcal{E}_p$. Let $(\mathcal{E}_p) = \{\varphi \in (L^2); \ \|\varphi\|_p < \infty\}$. The projective limit $(\mathcal{E})$ of $\{(\mathcal{E}_p); \ p \geq 1\}$ serves as a space of test functions on the abstract white noise space $(\mathcal{E}', \mu)$. The dual space $(\mathcal{E})^*$ of $(\mathcal{E})$ serves as the corresponding space of generalized functions. Thus we have the Gel'fand triple

$$(\mathcal{E}) \hookrightarrow (L^2) \hookrightarrow (\mathcal{E})^*.$$

The space $(\mathcal{E})$ of test functions is an infinite dimensional analogue of the Schwartz space $\mathcal{S}(\mathbb{R}^d)$ on the finite dimensional space $\mathbb{R}^d$ and has many similar properties as $\mathcal{S}(\mathbb{R}^d)$. For example, it is closed under pointwise multiplication of two test functions. The differential operators, translation operators, scaling operators, Fourier-Gauss transform, Gross Laplacian, and number operator are all continuous linear operators on $(\mathcal{E})$. Hence their adjoint operators are continuous on the dual space $(\mathcal{E})^*$. The integral kernel operators can be defined and are continuous linear operators from $(\mathcal{E})$ into $(\mathcal{E})^*$. Moreover, the Hitsuda-Skorokhod integral can be defined as a random variable on the white noise space $\mathcal{S}'(\mathbb{R})$.

4. Characterization theorems of Potthoff-Streit and Lee (1991)

In 1991 two important theorems were obtained by Potthoff and Streit [420] (for generalized functions) and by Y.-J. Lee [328] (for test functions).

The Potthoff-Streit theorem characterizes generalized functions in the space $(\mathcal{E})^*$ in terms of their S-transform

$$S\Phi(\xi) = \langle\!\langle \Phi, :e^{\langle \cdot, \xi \rangle}: \rangle\!\rangle, \quad \xi \in \mathcal{E}_c,$$

where $\mathcal{E}_c$ is the complexification of $\mathcal{E}$. The theorem says that a complex-valued function F on $\mathcal{E}_c$ is the S-transform of a generalized function in $(\mathcal{E})^*$ if and only if it satisfies the following conditions:

(a) For any $\xi, \eta \in \mathcal{E}_c$, the function $F(z\xi + \eta)$ is an entire function of $z \in \mathbb{C}$;

(b) There exist constants $K, a, p > 0$ such that

$$|F(\xi)| \leq K \exp\left[a|\xi|_p^2\right], \quad \forall \xi \in \mathcal{E}_c.$$

The Lee theorem describes the space $(\mathcal{E})$ of test functions directly in terms of analyticity and growth condition. It says that a function φ on $\mathcal{E}'$ belongs to $(\mathcal{E})$ if and only if for any $p \geq 1$ the function φ is analytic on $\mathcal{E}'_{p,c}$ (the complexification of $\mathcal{E}'_p$) and there exists a constant $K_p \geq 0$ such that

$$|\varphi(x)| \leq K_p \exp\left[\frac{1}{2}|x|^2_{-p}\right], \quad \forall x \in \mathcal{E}'_{p,c}.$$

On the other hand, test functions in $(\mathcal{E})$ can also be characterized in terms of their S-transform [304]. A complex-valued function F on $\mathcal{E}_c$ is the S-transform of a test function in $(\mathcal{E})$ if and only if it satisfies the following conditions:

(a) For any $\xi, \eta \in \mathcal{E}_c$, the function $F(z\xi + \eta)$ is an entire function of $z \in \mathbb{C}$;

(b) For any $a, p > 0$, there exists a constant $K > 0$ such that

$$|F(\xi)| \leq K \exp\left[a|\xi|^2_{-p}\right], \quad \forall \xi \in \mathcal{E}_c.$$

5. More spaces of generalized functions (1992–2000)

In 1992 Kondratiev and Streit introduced a family of Gel'fand triples [243] [244]. Let $0 \leq \beta < 1$. For φ represented by Equation (3.1) and $p \geq 1$, define

$$\|\varphi\|_{p,\beta} = \left(\sum_{n=0}^{\infty} (n!)^{1+\beta} |f_n|_p^2 \right)^{1/2}.$$

Let $(\mathcal{E}_p)_\beta = \{\varphi \in (L^2); \|\varphi\|_{p,\beta} < \infty\}$. The projective limit $(\mathcal{E})_\beta$ of $\{(\mathcal{E}_{p,\beta}); p \geq 1\}$ and its dual space $(\mathcal{E})_\beta^*$ give a new Gel'fand triple

$$(\mathcal{E})_\beta \hookrightarrow (L^2) \hookrightarrow (\mathcal{E})_\beta^*.$$

Kondratiev and Streit showed in [243] [244] that generalized functions in $(\mathcal{E})_\beta^*$ and test functions in $(\mathcal{E})_\beta$ can be characterized as follows. A complex-valued function F on $\mathcal{E}_c$ is the S-transform of a generalized function in $(\mathcal{E})_\beta^*$ if and only if it satisfies the conditions:

(a) For any $\xi, \eta \in \mathcal{E}_c$, the function $F(z\xi + \eta)$ is an entire function of $z \in \mathbb{C}$;

(b) There exist constants $K, a, p > 0$ such that

$$|F(\xi)| \leq K \exp\left[a|\xi|_p^{\frac{2}{1-\beta}}\right], \quad \forall \xi \in \mathcal{E}_c. \tag{5.1}$$

On the other hand, a complex-valued function F on $\mathcal{E}_c$ is the S-transform of a test function in $(\mathcal{E})_\beta$ if and only if it satisfies the conditions:

(a) For any $\xi, \eta \in \mathcal{E}_c$, the function $F(z\xi + \eta)$ is an entire function of $z \in \mathbb{C}$;

(b) For any $a, p > 0$, there exists a constant $K > 0$ such that

$$|F(\xi)| \leq K \exp\left[a|\xi|_{-p}^{\frac{2}{1+\beta}}\right], \quad \forall \xi \in \mathcal{E}_c. \tag{5.2}$$

Test functions in the space $(\mathcal{E})_\beta$ have also been described in [290] in the same spirit of Y.-J. Lee, namely, a function φ on $\mathcal{E}'$ belongs to $(\mathcal{E})_\beta$ if and only if for any $p \geq 1$ the function φ is analytic on $\mathcal{E}'_{p,c}$ and there exists a constant $K_p \geq 0$ such that

$$|\varphi(x)| \leq K_p \exp\left[\frac{1}{2}(1+\beta)|x|_{-p}^2\right], \quad \forall x \in \mathcal{E}'_{p,c}.$$

The collection $(\mathcal{E})_\beta^*$, $0 \leq \beta < 1$, is an increasing family of generalized functions. Kondratiev and Streit [244] also constructed a space $(\mathcal{E})^{-1}$ of generalized functions dealing with the case $\beta = 1$. But the corresponding space of test functions is not a nuclear space.

Between the union $\cup_{0\leq\beta<1}(\mathcal{E})^*_\beta$ and the space $(\mathcal{E})^{-1}$ there is a huge gap of generalized functions. This gap is filled up by the construction of Cochran et al. [82] associated with a sequence of positive real numbers $\{\alpha(n)\}_{n=0}^\infty$ satisfying the conditions:

(1) $\alpha(0) = 1$, $\inf_{n\geq 0} \alpha(n)\sigma^n > 0$ for some constant $\sigma \geq 1$.

(2) $\lim_{n\to\infty} \left(\frac{\alpha(n)}{n!}\right)^{1/n} = 0$.

For φ represented by Equation (3.1) and $p \geq 1$, define

$$\|\varphi\|_{p,\alpha} = \left(\sum_{n=0}^\infty n!\alpha(n)|f_n|_p^2\right)^{1/2}.$$

Let $[\mathcal{E}_p]_\alpha = \{\varphi \in (L^2); \|\varphi\|_{p,\alpha} < \infty\}$ and define $[\mathcal{E}]_\alpha$ to be the projective limit of $\{[\mathcal{E}_{p,\alpha}]; p \geq 1\}$. The dual space $[\mathcal{E}]^*_\alpha$ is a new space of generalized functions and we have the Gel'fand triple

$$[\mathcal{E}]_\alpha \hookrightarrow (L^2) \hookrightarrow [\mathcal{E}]^*_\alpha.$$

Important examples of $\{\alpha(n)\}$ given by the Bell numbers [82]. A characterization theorem for generalized functions in the space $[\mathcal{E}]^*_\alpha$ is given in [82] with the growth condition: there exist constants $K, a, p > 0$ such that

$$|F(\xi)| \leq KG_\alpha\left(a|\xi|_p^2\right)^{1/2}, \tag{5.3}$$

where $G_\alpha(r) = \sum_{n=0}^\infty \frac{\alpha(n)}{n!}r^n$. The corresponding characterization theorem for test functions in the space $[\mathcal{E}]_\alpha$ is given in [36] with the growth condition: for any $a, p > 0$, there exists a constant $K \geq 0$ such that

$$|F(\xi)| \leq KG_{1/\alpha}\left(a|\xi|_{-p}^2\right)^{1/2}, \tag{5.4}$$

where $G_{1/\alpha}(r) = \sum_{n=0}^\infty \frac{1}{\alpha(n)n!}r^n$.

For the special case $\alpha(n) = (n!)^\beta$, the growth functions in Equations (5.3) and (5.4) are different from those in Equations (5.1) and (5.2), respectively. Moreover, for a given sequence $\{\alpha(n)\}$, it is impossible in general to sum up G_α and $G_{1/\alpha}$ in close forms. This is the motivation for Asai et al. [33] [35] [38] [39] to introduce CKS-space associated with a growth function.

Let u be a continuous function on $[0, \infty)$ satisfying the following conditions:

(1) $\lim_{r\to\infty} \frac{\log u(r)}{\sqrt{r}} = \infty$.

(2) $\limsup_{r\to\infty} \frac{\log u(r)}{r} < \infty$.

(3) $u(0) = 1$ and u is increasing.

(4) $\log u(x^2)$ is convex for $x \in [0, \infty)$.

Define the Legendre transform ℓ_u of u by

$$\ell_u(n) = \inf_{r>0} \frac{u(r)}{r^n}, \qquad n = 0, 1, \ldots,$$

and the dual Legendre transform u^* of u by

$$u^*(r) = \sup_{s \geq 0} \frac{e^{2\sqrt{rs}}}{u(s)}, \qquad r \geq 0.$$

For such a function u, define a sequence $\alpha_u(n)$ by

$$\alpha_u(n) = \frac{1}{\ell_u(n)n!}, \qquad n = 0, 1, \ldots.$$

With this sequence we get a CKS-space denoted by

$$[\mathcal{E}]_u \hookrightarrow (L^2) \hookrightarrow [\mathcal{E}]_u^*.$$

Characterization theorems for generalized functions in $[\mathcal{E}]_u^*$ and test functions in $[\mathcal{E}]_u$ are given in [39]. A complex-valued function F on $\mathcal{E}_c$ is the S-transform of a generalized function in $[\mathcal{E}]_u^*$ if and only if it satisfies the conditions:

(a) For any $\xi, \eta \in \mathcal{E}_c$, the function $F(z\xi + \eta)$ is an entire function of $z \in \mathbb{C}$;

(b) There exist constants $K, a, p > 0$ such that

$$|F(\xi)| \leq K u^*\big(a|\xi|_p^2\big)^{1/2}, \quad \forall \xi \in \mathcal{E}_c.$$

On the other hand, a complex-valued function F on $\mathcal{E}_c$ is the S-transform of a test function in $[\mathcal{E}]_u$ if and only if it satisfies the conditions:

(a) For any $\xi, \eta \in \mathcal{E}_c$, the function $F(z\xi + \eta)$ is an entire function of $z \in \mathbb{C}$;

(b) For any $a, p > 0$, there exists a constant $K > 0$ such that

$$|F(\xi)| \leq K u\big(a|\xi|_{-p}^2\big)^{1/2}, \quad \forall \xi \in \mathcal{E}_c.$$

6. Applications of white noise theory

There is a wide range of applications of white noise theory. We mention just a few of them below with very brief comments.

1. Integral kernel operators

In 1992 Hida, Obata, and Saitô introduced the integral kernel operators [193]. The generalization to operator-valued generalized functions has been done by Obata, Chung, and Ji, among others. These operators are important for applications in quantum probability.

2. Stochastic integration

In 1981 Kubo and Takenaka related the white noise integral $\int_a^b \partial_t^* \varphi(t)\, dt$ to the Itô integral $\int_a^b \varphi(t)\, dB(t)$. The integral $\int_a^b \partial_t^* \varphi(t)\, dt$ turns out to be the same as the integral introduced by Hitsuda in 1972 [206] and by Skorokhod in 1975 [472]. In the paper [247] Kubo introduced the Itô formula for $f(B(t))$ with f being a generalized function. In 1990 Kuo and Potthoff [302] [303] studied anticipating stochastic integrals and stochastic differential equations. For more information, see the book [290]. Recent progress on white noise methods for stochastic integration has been done by Potthoff and Streit and their collaborators.

3. Infinite dimensional harmonic analysis

In 1975 Hida [123] outlined several things about the infinite dimensional rotation group on the white noise space. Through the years the subgroups of this group, in particular the Lévy group, have been analyzed very much by Hida, Obata, and Saitô. The Lévy Laplacian and related semigroups have been studied extensively by Saitô and his collaborators. Recently Lee and Stan [344] obtained the infinite dimensional Heisenberg inequality. The finite dimensional Paley-Wiener theorem has been generalized to the white noise space by Stan in [473].

4. Stochastic partial differential equations

Since 1991 Øksendal and his collaborators have made significant progress using white noise theory to study stochastic partial differential equations. See the book [217]. Recent achievements of white noise methods to study SPDE's (in particular the Burgers equation) have been made by Øksendal, Potthoff, Streit, and their collaborators [49] [50] [53] [87] [88] [90] [109] [212] [213].

5. Feynman integral

Feynman integral was one of the motivations for Hida to introduce white noise theory in 1975. Streit and his numerous collaborators have made revolutionary work to treat Feynman integrands as generalized functions in various spaces of generalized functions. In the paper [267] it is shown that with a rather general potential function the associated Feynman integrand is a generalized function in some big space of generalized functions. Recently, Asai et al. [17] [18] have shown that for the Albeverio–Høegh-Krohn and Laplace transform potentials the associated Feynman integrands are generalized functions in the space $[\mathcal{E}]_u^*$ for some growth function u.

6. Random fields and stochastic variational calculus

Since 1988 Hida and Si Si have made significant progress for this application of white noise theory. For further information, see the new book by Hida [184].

7. Intersection local times

In 1991 H. Watanabe [500] used white noise theory to study the local time of self-intersections of Brownian motions. Recently Streit and his collaborators have obtained rather interesting results in [101].

8. Infinite dimensional stochastic differential equations

White noise theory can be used to study the laws of solutions of $\mathcal{S}'(\mathbb{R})$-valued stochastic differential equations. In [311] it is shown that under certain conditions the laws of the solution of an $\mathcal{S}'(\mathbb{R})$-valued SDE induce generalized functions in the space $(\mathcal{S})^*$ of generalized functions. Moreover, these generalized functions satisfy an infinite dimensional partial differential equations.

9. Positive generalized functions

Positive generalized functions in white noise theory was first studied by Potthoff [409] in 1987. These functions are associated with Hida measures on a white noise space. Interesting results have been obtained by Yokoi, Y.-J. Lee, Asai-Kubo-Kuo, on various spaces of generalized functions. For Hida measures on the Kondratiev-Streit space, see the book [290].

10. Dirichlet forms

In 1988 Hida, Potthoff, and Streit [195] initiated the application of white noise theory to Dirichlet forms. Later they were joined by Albeverio and Röckner [23] [24] to achieve revolutionary work on the construction of Dirichlet forms for quantum field theory.

11. Quantum probability

Recent development of quantum probability by Accardi and his collaborators (in particular, Lu, Obata, and Volovich) has shown a strong connection between white noise theory and quantum probability. See the IIAS Report [16] and the Volterra Center Preprint [17] and the references therein.

7. References on white noise theory

In the references below we have compiled a list of publications on white noise theory since 1975 with a few exceptions, which are listed for their influence on white noise theory. This list is by no means complete, but it gives some ideas on the development of white noise theory since 1975.

In addition, there have been several conference proceedings and special volumes which contain many papers related to white noise theory:

(1) *White Noise Analysis: Math. and Appls.*, T. Hida, H.-H. Kuo, J. Potthoff, and L. Streit (eds.), World Scientific, Singapore, 1990

(2) *Gaussian Random Fields*, K. Itô and T. Hida (eds.), World Scientific, Singapore, 1991

(3) *Stochastic Analysis on Infinite Dimensional Spaces*, H. Kunita and H.-H. Kuo (eds.), Longman Science & Technical, Pitman Research Notes in Math. Series, vol. 310, 1994

(4) *Trends in Contemporary Infinite Dimensional Analysis and Quantum Probability*, L. Accardi, N. Obata, K. Saitô, Si Si, and L. Streit (eds.), Italian School of East Asian Studies, Natural and Mathematical Sciences Series 3, Intituto Italiano di Cultura, Kyoto, 2000

(5) *Recent Developments in Infinite Dimensional Analysis and Quantum Probability*, in honor of T. Hida's 70th birthday, L. Accardi, N. Obata, K. Saitô, Si Si, and L. Streit (eds.), a special issue in *Acta Applicandae Mathematicae*, vol. 63, nos. 1–3, 2000

(6) *Collected Papers of T. Hida*, L. Accardi, N. Obata, K. Saitô, Si Si, and L. Streit (eds.), World Scientific, 2001

Acknowledgements. (1) This article was prepared during my visit to Hiroshima University, May 20–August 19, 2001. I would like to give my deepest appreciation to Professor I. Kubo for submitting a proposal to Monbu-Kagaku-Sho (Ministry of Education and Science) for my visit to him and Hiroshima University. I am most grateful for the warm hospitality of the mathematics department and the personnel office of the Graduate School of Science, Hiroshima University. I want to give my best thanks to Monbu-Kagaku-Sho for the financial support for my visit to Hiroshima University.

(2) I would like to thank the Academic Frontier in Science of Meijo University for financial supports and to give my deep appreciation to Professors T. Hida and K. Saitô for the warm hospitality during my many visits over the years 1998–2001.

References

[1] Aase, K., Øksendal, B., Privault, N., and Ubøe, J.: A white noise generalization of the Clark-Haussmann-Ocone theorem, with application to mathematical finance; *Finance and Stochastics* **4** (2000) 465–496

[2] Aase, K., Øksendal, B., and Ubøe, J.: Using the Donsker delta function to compute hedging strategies; *Preprint University of Oslo* 6/1998 and *Potential Analysis* (to appear)

[3] Accardi, L.: Quantum stochastic calculus; *Probability Theory and Mathematical Statistics* 1 (1985) 1–21

[4] Accardi, L.: Yang-Mills equations and Lévy-Laplacian; in: *Dirichlet Forms and Stochastic Processes*, de Gruyter (1995) 1–24

[5] Accardi, L. and Bogachev, V.: The Ornstein-Uhlenbeck process and the Dirichlet form associated to the Lévy Laplacian; *C. R. Acad. Sci. Paris Ser. I Math.* bf 320 (1995) 597–602

[6] Accardi, L. and Bogachev, V.: The Ornstein-Uhlenbeck process associated with the Lévy Laplacian and its Dirichlet form; *Probab. Math. Statist.* 17 (1997) 95–114

[7] Accardi, L. and Bogachev, V.: On stochastic analysis associated with the Lévy Laplacian; *Dokl. Akad. Nauk* 358 (1998) 583–587

[8] Accardi, L., Boukas, A., and Kuo, H.-H.: On the unitarity of stochastic evolutions driven by the square of white noise; *Preprint* (2000)

[9] Accardi, L. and Bożejko, M.: Interacting Fock space and Gaussianization of probability measures; *Infinite Dimensional Analysis, Quantum Probability and Related Topics* 1 (1998) 663–670

[10] Accardi, L., Gibilisco, P., and Volovich, I. V.: Yang-Mills gauge fields as harmonic functions for the Lévy Laplacian; *Russian J. Math. Phys.* 2 (1994) 235–250

[11] Accardi, L., Hashimoto, Y., and Obata, N..: Notions of independence related to the free group; *Infinite Dim. Anal., Quantum Prob. and Related Topics* 1 (1998) 201–220

[12] Accardi, L., Hashimoto, Y., and Obata, N..: Singleton independence; *Banach Center Publications* 43 (1998) 9–24

[13] Accardi, L., Hashimoto, Y., and Obata, N..: A role of singletons in quantum central limit theorems; *J. Korean Math. Soc.* 35 (1998) 675–690

[14] Accardi, L., Hida, T., and Kuo, H.-H.: Itô table for the square of white noise; *Preprint* (2000)

[15] Accardi, L., Hida, T., and Win Win Htay: Boson Fock representations of stationary processes (in Russian); *Mathematical Notes* 67 (2000) 3–14

[16] Accardi, L., Lu, Y. G., and Volovich, I.: The Hilbert module and interacting Fock spaces; *IIAS Reports* 1997-008 (1997)

[17] Accardi, L., Lu, Y. G., and Volovich, I. V.: White noise approach to classical and quantum stochastic calculi; *Centro Vito Volterra, Università di Roma "Tor Vergata"* 375 (1999)

[18] Accardi, L., Lu, Y. G., and Volovich, I. V.: A white noise approach to stochastic calculus; *Acta Appl. Math.* 63 (2000) 3-25

[19] Accardi, L., Rozelli, P., and Smolyanov, O. G.: Brownian motion generated by the Lévy Laplacian; *Mat. Zametki* 54 (1993) 144–148

[20] Albeverio, S., Daletzky, Y., Kondratiev, Yu. G., and Streit, L.: Non-Gaussian infinite dimensional analysis; *J. Funct. Analysis* 138 (1996) 311–350

[21] Albeverio, Ø., Fukushima, M., Karwowski, W., and Streit, L.: Capacity and quantum mechanical tunneling; *Comm. math. Phys* 81 (1981) 501

[22] Albeverio, S., Gesztesy, F., Karwowski, W., and Streit, L.: On the connection between Schrödinger and Dirichlet forms; *J. Math. Phys* **26** (1985) 2546

[23] Albeverio, S., Hida, T., Potthoff, J., Röckner, M., and Streit, L.: Dirichlet forms in terms of white noise analysis I–Construction and QFT examples; *Rev. Math. Phys.* **1** (1990) 291–312

[24] Albeverio, S., Hida, T., Potthoff, J., Röckner, M., and Streit, L.: Dirichlet forms in terms of white noise analysis II–Closability and diffusion processes; *Rev. Math. Phys.* **1** (1990) 313–323

[25] Albeverio, S., Hida, T., Potthoff, J., and Streit, L.: The Vacuum of the Hoegh-Krohn model as a generalized white noise functional; *Phys. Lett. B* **217** (1989) 511–514

[26] Albeverio, S., Hoegh-Krohn, R., and Streit, L.: Energy forms, Hamiltonians, and distorted Brownian paths; *J. Math. Phys.* **18** (1977) 907

[27] Albeverio, S., Karwowski, W., Röckner, M., and Streit, L.: Capacity, Green's functions and Schroedinger equation; in: *Infinite Dimensional Analysis and Stochastic Processes*, Pitman 1985

[28] Albeverio, S., Kondratiev, Yu. G., and Streit, L.: How to generalize white noise analysis to non-Gaussian measures; in: *Dynamics of Complex and Irregular Systems*, Ph. Blanchard et al. (eds.), World Scientific (1993) 120–130

[29] Asai, N:: A note on general setting of white noise analysis and positive generalized functions; *RIMS Kokyuroku* **1139** (2000) 19–29

[30] Asai, N: Characterization of product measures by integrability condition; to appear in: Quantum Information III, T. Hida and K. Saitô (eds.) World Scientific (2001) 21–33

[31] Asai, N: Analytic characterization of one-mode interacting Fock space; *Infinite Dimensional Analysis, Quantum Probability and Related Topics* (to appear)

[32] Asai, N: Integral transform and Segal-Bargmann representation associated to q-Charlier polynomials; *Quantum Information* **4**, World Scientific (2001)

[33] Asai, N., Kubo, I., and Kuo, H.-H.: Characterization of Hida measures in white noise analysis; in *Infinite Dimensional Harmonic Analysis, Transactions of a Japan-Germany Symposium*, H. Heyer et al. (eds.) (1999) 70–83, Druck D.+M. Gräbner

[34] Asai, N., Kubo, I., and Kuo, H.-H.: Feynman integrals associated with Albeverio–Høegh-Krohn and Laplace transform potentials; in *Stochastic in Finite and Infinite Dimensions, in honor of G. Kallianpur*, T. Hida et al. (eds.) (2000) 29–48, Birkhäuser

[35] Asai, N., Kubo, I., and Kuo, H.-H.: CKS-space in terms of growth functions; in *Quantum Information II*, T. Hida and K. Saitô (eds.) (2000) 17–27, World Scientific

[36] Asai, N., Kubo, I., and Kuo, H.-H.: Characterization of test functions in CKS-space; in *Mathematical Physics and Stochastic Analysis: Essays in Honor of Ludwig Streit*, A. Albeverio et al. (eds.) (2000) 68–78, World Scientific

[37] Asai, N., Kubo, I., and Kuo, H.-H.: Bell numbers, log-concavity, and log-convexity; *Acta Applicandae Mathematicae* **63** (2000) 79–87

[38] Asai, N., Kubo, I., and Kuo, H.-H.: Roles of log-concavity, log-convexity, and growth order in white noise analysis; *Infinite Dimensional Analysis, Quantum Probability, and Related Topics* **4** (2001) 59–84

[39] Asai, N., Kubo, I., and Kuo, H.-H.: General characterization theorems and intrinsic topologies in white noise analysis; *Hiroshima Math. J.* **31** (2001) 299–330

[40] Asai, N., Kubo, I., and Kuo, H.-H.: Segal-Bargmann transforms of one-mode interacting Fock spaces associated with Gaussian and Poisson measures; *Preprint* (2001)

[41] Asch, J. and Potthoff, J.: A generalization of Itô's lemma; *Proc. Japan Acad.* **63** (1987) 289–291

[42] Asch, J. and Potthoff, J.: White noise and stochastic analysis; in: *Stochastics, Algebra and Analysis in Classical and Quantum Dynamics*, S. Albeverio et al. (eds.), Kluwer Academic Publishers (1990)

[43] Asch, J. and Potthoff, J.: Itô's lemma without non-anticipatory conditions; *Probab. Th. Rel. Fields* **88** (1991) 17–46

[44] Benth, F. E.: Integrals in the Hida distribution space $(\mathcal{S})^*$; *Stochastic Monogr.* **8** (1993) 89–99

[45] Benth, F. E.: A generalized Feynman-Kac formula for the stochastic heat problem with anticipating initial conditions; *Progr. Probab.* **38** (1996) 121–133

[46] Benth, F. E.: On the positivity of the stochastic heat equation; *Potential Anal.* **6** (1997) 127–148

[47] Benth, F. E.: The Gross derivative and generalized random variables; *Infinite Dimensional Analysis, Quantum Probability and Related Topics* bf 2 (1999) 381–396

[48] Benth, F., Deck, Th., Potthoff, J.: A white noise approach to a class of non-linear stochastic heat equations; *J. Funct. Anal.* **146** (1997) 382–415

[49] Benth, F. E., Deck, Th., Potthoff, J., and Streit, L.: Nonlinear evolution equations with gradient coupled noise; *Lett. Math. Phys.* **43** (1998) 267–278

[50] Benth, F. E., Deck, Th., Potthoff, J, and Vage, G.: Explicit strong solutions of SPDE's with applications to non-linear filtering; *Acta Appl. Math.* **51** (1998) 215–242

[51] Benth, F. E., Øksendal, B., Ubøe, J., and Zhang, T.: Wick products of complex valued random variables; in: *Stochastic Analysis and Related Topics (5)*, H. Korezlioglu et al. (eds.), Birkhäuser (1996) 135–155

[52] Benth, F. and Potthoff, J.: On the martingale property for generalized stochastic processes; *Stochastics and Stochastics Reports* **58** (1996) 349–367

[53] Benth, F. E. and Streit, L.: The Burgers equation with a non-Gaussian random force; *Progr. Probab.* **42** (1998) 187–210, Birkhäuser

[54] Bernido, C. C. and Carpio-Bernido, M.V.: Path integrals for boundaries and topological constraints: A white noise functional approach; *Preprint* (2001)

[55] Betounes, D.: Generalized stochastic differential equations on $(D)^*$; *Lecture Notes in Control and Information Sciences* **176** (1992) 32–37

[56] Betounes, D.: Trace operators, Feynman distributions, and multiparameter white noise; *Journal of Theoretical Probability* **8** (1995) 119–138

[57] Betounes, D. and Redfern, M.: Differential forms with values in $(L^2)^-$; in: *White Noise Analysis*, World Scientific (1990) 31–42

[58] Betounes, D. and Redfern, M.: Wiener distributions and white noise analysis; *J. Appl. Math. Optim.* **26** (1992) 63–93

[59] Betounes, D. and Redfern, M.: Stochastic integrals for nonprevisible, multiparameter processes; *J. of Appl. Math. Optim.* **28** (1993) 197–223

[60] Betounes, D. and Redfern, M.: The Stratonovich integral via the renormalization operator on Fock space; *Stochastics and Stochastics Reports* **56** (1996) 161–178

[61] Blümlinger, M. and Obata, N.: Permutations preserving Cesàro mean, densities of natural numbers and uniform distribution of sequences; *Ann. Inst. Fourier Grenoble* **41** (1991) 665–678

[62] Carmona, R. and Yan, J. A.: A new space of white noise distributions and applications to SPDE's; in: *Seminar on Stochastic Analysis, Random Fields and Applications* (1995) 51–66

[63] Chow, P. L.: Generalized solution of some parabolic equations with a random drift; *J. Appl. Math. Optim.* **20** (1989) 81–96

[64] Chow, P. L.: Stationary solutions of some parabolic Itô equations; *Pitman Research Notes in Math. Series* **310** (1994) 42–51, Longman Scientific & Technical

[65] Chung, D. M. and Chung, T. S.: First order differential operators in white noise; *Proc. Amer. Math. Soc.* **126** (1998) 2367–2376

[66] Chung, D. M., Chung, T. S., and Ji, U. C.: A simple proof of analytic characterization theorem for operator symbols; *Bull. Korean Math. Soc.* **34** (1997) 421–436

[67] Chung, D. M., Chung, T. S., and Ji, U. C.: Products of white noise functionals and associated derivations; *J. Korean Math. Soc.* **35** (1998) 559–574

[68] Chung, D. M., Chung, T. S., and Ji, U. C.: A characterization theorem for operators on white noise functionals; *J. of Math. Soc. Japan* **51** (1999) 437–447

[69] Chung, D. M. and Ji, U. C.: Cauchy Problems for a Partial Differential Equation in White Noise Analysis; *J. Korean Math. Soc.* **33** (1996) 309–318

[70] Chung, D. M. and Ji, U. C.: Wick Derivations on White Noise Functionals; *J. Korean Math. Soc.* **33** (1996) 559–574

[71] Chung, D. M. and Ji, U. C.: Transforms on white noise functionals with their applications to Cauchy problems; *Nagoya Math. J.* **147** (1997) 1–23

[72] Chung, D. M. and Ji, U. C.: Transformation groups on white noise functionals and their applications; *Appl. Math. Optim.* **37** (1998) 205–223

[73] Chung, D. M. and Ji, U. C.: Some Cauchy problems in white noise analysis and associated semigroups of operators; *Stoch. Analy. Appl.* **17** (1999) 1–22

[74] Chung, D. M. and Ji, U. C.: Multi-parameter transformation groups on white noise functionals; *J. Math. Anal. and Appl.* **252** (2000) 729–749

[75] Chung, D. M. and Ji, U. C.: Poisson equations associated with differential second quantization operators in white noise analysis; *Acta Applicadae Mathematicae* **63** (2000) 89–100

[76] Chung, D. M., Ji, U. C, and Obata, N.: Transformations on white noise functions associated with second order differential operators of diagonal type; *Nagoya Math. J.* **149** (1998) 173–192

[77] Chung, D. M., Ji, U. C, and Obata, N.: Higher powers of quantum white noises in terms of integral kernel operators; *Infinite Dim. Anal., Quantum Prob. and Related Topics* **1** (1998) 533–559

[78] Chung, D. M., Ji, U. C, and Obata, N.: Normal-ordered white noise differential equations II: Regularity properties of solutions; in: *Probability Theory and Mathematical Statistics*, B. Grigolionis et al. (eds), VSP BV and TEV Ltd. (1999) 157–174

[79] Chung, D. M., Ji, U. C, and Obata, N.: Normal-ordered white noise differential equations I: Existence of solutions as Fock space operators; in: *Trends in Contemporary Infinite Dimensional Analysis and Quantum Probability*, L. Accardi et al. (eds), Istituto Italiano di Cultura, Kyoto (2000) 115–135

[80] Chung, D. M., Ji, U. C., and Saitô, K.: Cauchy problems associated with the Lévy Laplacian in white noise analysis; *Infinite Dimensional Analysis, Quantum Probability and Related Topics* **2** (1999) 131–153

[81] Chung, D. M., Ji, U. C., and Saitô, K.: Notes on a C_0-group generated by the Lévy Laplacian; *Proc. American Mathematical Society* (2001)

[82] Cochran, W. G., Kuo, H.-H., and Sengupta, A.: A new class of white noise generalized functions; *Infinite Dimensional Analysis, Quantum Probability, and Related Topics* **1** (1998) 43–67

[83] Cochran, W. G., Lee, J.-S., and Potthoff, J.: Stochastic Volterra equations with singular kernels; *Stoch. Proc. Appl.* **56** (1995) 337–349

[84] Cochran, W. G. and Potthoff, J.: Fixed point principles for stochastic partial differential equations; in: *Dynamics of Complex and Irregular Systems*, Ph. Blanchard et al. (eds.), World Scientific (1993)

[85] Deck, Th.: Hida distributions over compact Lie groups; *Infinite Dim. Anal. Quantum Probab. and Related Topics* **3** (2000) 337–362

[86] Deck, Th.: Continuous dependence on initial data for non-linear stochastic evolution equations; *Preprint* (2001)

[87] Deck, Th., Kruse, S., Potthoff, J., and Watanabe, H.: White noise approach to stochastic partial differential equations; in: *Proc. Int. Conf. on SPDE's*, Levico (2000)

[88] Deck, Th. and Potthoff, J.: On a class of stochastic partial differential equations related to turbulent transport; *Probab. Th. Rel. Fields* **111** (1998) 101–122

[89] Deck, Th., Potthoff, J., and Vage, G.: A review of white noise analysis from a probabilistic standpoint; *Acta Appl. Math.* **48** (1997) 91–112

[90] Deck, Th., Potthoff, J., Vage, G., and Watanabe, H.: Stability of solutions of parabolic PDE's with random drift and viscosity limit; *Appl. Math. Optim.* **40** (1999) 393–406

[91] de Falco, D. and Khandekar, D. C.: Applications of white noise calculus to the computation of Feynman integrals; *Stochastic Processes and Their Applications* **29** (1988) 257-266

[92] de Faria, M., Drumond, C., and Streit, L.: The renormalization of self-intersection local times. I. The chaos expansion; *Infinite Dimensional Analysis, Quantum Probability and Related Topics* **3** (2000) 223–236

[93] de Faria, M., Hida, T., Streit, L., and Watanabe, H.: Intersection local times as generalized white noise functionals; *Acta Applicandae Mathematicae* **46** (1997) 351–362

[94] de Faria, M. and Kuo, H.-H.: A delta white noise functional; *Acta Applicandae Mathematicae* **17** (1989) 287–298

[95] de Faria, M., Oliveira, M. J., and Streit, L.: A generalized Clark-Ocone formula; *Random Oper. Stoch. Eqs.* **8** (2000) 163–174

[96] de Faria, M., Potthoff, J., and Streit, L.: The Feynman integrand as a Hida distribution; *J. Math. Phys.* **32** (1991) 2123–2127

[97] de Faria, M. and Streit, L.: Some recent advances in white noise analysis; *Pitman Research Notes in Math. Series* **310** (1994) 52–59, Longman Scientific & Technical

[98] Dôku, I.: On a class of infinite-dimensional Fourier type transforms in white noise calculus; in: *Probability Theory and Mathematical Statistics*, World Scientific (1995) 51–61

[99] Dôku, I.: On the Laplacian on a space of white noise functionals; *Tsukuba J. Math.* **19** (1995) 93–119

[100] Dôku, I., Kuo, H.-H., and Lee, Y.-J.: Fourier transform and heat equation in white noise analysis; *Pitman Research Notes in Math. Series* **310** (1994) 60–74, Longman Scientific & Technical

[101] Drumond, C., de Faria, M., and Streit, L.: The square of self-intersection local time of Brownian motion; in: *Stochastic Processes, Physics and Geometry: New Interplays I*, a volume in honor of Sergio Albeverio, F. Gesztesy et al. (eds.) - CMS Conf. Proc. 28, Amer. Math Soc. (2000) 115–122

[102] Gannoun, R., Hachaichi, R., Ouerdiane, H., and Rezgui, A.: Un théorème de dualité entre espaces de fonctions holomorphes à croissance exponentiele; *J. Funct. Anal.* **171** (2000) 1–14

[103] Gielerak, R., Karwowski, W., and Streit, L.: Construction of a class of characteristic functionals; in: *Feynman Path Integrals*, S. Albeverio et al. (eds.), *Lecture Notes in Physics* **106** (1979) 182–188, Springer

[104] Gjerde, J., Holden, H., Øksendal, B., Ubøe, J., and Zhang, T.: An equation modelling transport of a substance in a stochastic medium; in: *Seminar on Stochastic Analysis, Random Fields and Applications*, E. Bolthausen et al. (eds.), Progress in Probability, Vol.36, Birkhäuser (1995) 123–134

[105] Gjessing, H., Holden, H., Lindstrøm, T., Øksendal, B., Ubøe, J., and Zhang, T.: The Wick product; in: *Frontiers in Pure and Applied Probability, Vol. I*, H. Niemi et al. (eds.), TVP Science Publishers, (1993) 29–67

[106] Grothaus, M., Khandekar, D. C., da Silva, J. L., and Streit, L.: The Feynman integral for time-dependent anharmonic oscillators; *J. Math. Phys.* **38** (1997) 3278–3299

[107] Grothaus, M., Kondratiev, Yu. G., and Streit, L.: Complex Gaussian analysis and the Bargmann-Segal space; *Methods of Funct. Anal. Topology* **3** (1997) 46–64

[108] Grothaus, M., Kondratiev, Yu. G., and Streit, L.: Regular generalized functions in Gaussian analysis; *Infinite Dim. Anal. Quantum Prob. and Related Topics* **2** (1999) 1–25

[109] Grothaus, M., Kondratiev, Yu. G., and Streit, L.: Scaling limits for the solution of Wick type Burgers equation; *Random Oper. Stochastic Equations* **8** (2000) 1–26

[110] Grothaus, M., Kondratiev, Yu. G., and Us, G. F.: Wick calculus for regular generalized stochastic functionals; *Random Oper. Stochastic Equations* **7** (1999) 263–290

[111] Grothaus, M. and Streit, L.: Construction of relativistic quantum fields in the framework of white noise analysis; *J. Math. Phys.* **40** (1999) 5387–5405

[112] Grothaus, M. and Streit, L.: Quadratic actions, semi-classical approximation, and delta sequences in Gaussian analysis; *Rep. Math. Phys.* **44** (1999) 381–405

[113] Grothaus, M. and Streit, L.: On regular generalized functions in white noise analysis and their applications; *Methods of Funct. Anal. Topology* **6** (2000) 14–27

[114] Grothaus, M., Streit, L., and Volovich, I.: Knots, Feynman diagrams and matrix models; *Infinite Dim. Anal. Quantum Prob. and Related Topics* **2** (1999) 359–380

[115] Hashimoto, Y., Obata, N., and Tabei, N.: A quantum aspect of asymptotic spectral analysis of large Hamming graphs; in: *Quantum Information III*, T. Hida and K. Saitô (eds.) (2001)

[116] Hida, T.: *Stationary Stochastic Processes.* Princeton University Press, 1970

[117] Hida, T.: Harmonic Analysis on the space of generalized functions; *Teor. Verojatnost. i Primenen.* (Theory of Probability and its applications) **15** (1970) 119-124

[118] Hida, T.: Note on the infinite dimensional Laplacian operator; *Nagoya Math. J.* **38** (1970) 13-19

[119] Hida, T.: Quadratic functional of Brownian motion; *J. Multivariate Analysis* **1** (1971) 58-69

[120] Hida, T.: *Complex white noise and infinite dimensional unitary group*, Lecture Notes, Mathematics Dept., Nagoya Univ. no.3, 1971

[121] Hida, T.: A role of Fourier transform in the theory of infinite dimensional unitary group; *J. Math. Kyoto Univ.* **13** (1973) 203–212.

[122] Hida, T.: Functionals of complex white noise; *Proc. Symp. on Continuous Mechanics and related problems of Analysis* **1** Tbilisi, USSR, 1973, 355-366

[123] Hida, T.: *Analysis of Brownian Functionals.* Carleton Mathematical Lecture Notes **13**, 1975

[124] Hida, T.: White noise analysis and nonlinear filtering problems; *Applied Mathematics and Optimization* **2** (1975) 82–89

[125] Hida, T.: Analysis of Brownian functionals; *Mathematical Programming Study* **5** (1976) 53–59

[126] Hida, T.: Functionals of Brownian motion; *Transaction of 7th Prague Conf. and 1974 European Meeting of Statisticians* **A** (1977) 239–243

[127] Hida, T.: Topics on nonlinear filtering theory; *Multivariate analysis* **4** (1977) 239–245

[128] Hida, T.: Generalized multiple Wiener integrals; *Proc. Japan Acad.* **54A** (1978) 55–58

[129] Hida, T.: Generalized Brownian functionals; *Complex Analysis and its Applications*, I. N. Vekua Volume (1978) 586–590

[130] Hida, T.: White noise and Levy's functional analysis; *Lecture Notes in Math.* **695** (1978) 155–163

[131] Hida, T.: White noise and Lévy's functional analysis; *Lecture Notes in Math.* **695** (1978) 155–163

[132] Hida, T.: Generalized multiple Wiener integrals; *Proc. Japan Acad.* **54A** (1978) 55–58

[133] Hida, T.: Nonlinear Brownian functionals; *Proc. 18th IEEE Conference on Decision and Control* **1** (1979) 326–328

[134] Hida, T.: *Brownian Motion*. Springer-Verlag, 1980

[135] Hida, T.: Causal analysis in terms of Brownian motion. Multivariate Analysis V,(ed. P. R. Krishnaiah) North Holland, 1980, 111-118.

[136] Hida, T.: Causal analysis in terms of white noise; in: *Quantum Fields - Algebra, Processes*, L. Streit (ed.) Springer-Verlag (1980) 1–19

[137] Hida, T.: Theory of probability, Gaussian processes and Physics (in Japanese); *Monthly Journal Physics* **2** (1980) 152–158

[138] Hida, T.: Causal calculus of Brownian functionals and its applications; *Proc. International Symposium on Statistics and Related Topics*, D.A. Dawson et al. (eds.) (1981) 353–360

[139] Hida, T.: White noise analysis and its applications; *Proc. International Mathematical Conference*, L. H. Y. Chen et al. (eds.) North Holland (1982) 43–48

[140] Hida, T.: Calculus of Brownian functionals; *Proc. International Mathematical Conference*, L. H. Y. Chen et al. (eds.) North Holland (1982) 155–185

[141] Hida, T.: The role of exponential functions in the analysis of generalized Brownian functionals; *Teor. Verojatnost. i Primenen.* **27** (1982) 569–573 (English translation *Theory of Prob. and its Appl.* **27** (1983) 609–613

[142] Hida, T.: Causal calculus and an application to prediction theory; *Prediction Theory and Harmonic Analysis*, V. Mandrekar and H. Salehi (eds.) (1983) 123–130, North-Holland

[143] Hida, T.: Generalized Brownian functionals; *Proc. IFIP-WG Theory and Application of Random Field, Lec. Notes in Control and Information Sci.* **49** (1983) 89–95

[144] Hida, T.: White noise analysis and its applications to quantum dynamics; *Physica* **124 A** (1984) 399–412

[145] Hida, T.: Generalized Brownian functionals and stochastic integrals; *Appl. Math. Optimization* **12** (1984) 115-123

[146] Hida, T.: Brownian motion and its functionals; *Ricerche di Matematica* **34** (1985) 183–222

[147] Hida, T.: Brownian functionals and the rotation group; *Mathematics + Physics*, L. Streit (ed.) **1** (1985) 167–194

[148] Hida, T.: White noise analysis and its applications to biology; *Proc. 15th NIBB Conf. Information Processing in Neuron Network*, K. Naka and Y. Ando (eds.) (1986) 3–13

[149] Hida, T.: Infinite dimensional rotation group and its applications to quantum dynamics; *Proc. 14th ICGTMP Conf.*, Y. M. Cho (ed.) (1986) 234–237

[150] Hida, T.: Generalized Gaussian measures; *Supplemento ai Rend. del Circolo Matematico Palermo, Ser. II* **17** (1987) 229–236

[151] Hida, T.: Generalized Gaussian measures; *Functional integrations with emphasis on the Feynman integrals*, Sherbrooks, PQ, 1986

[152] Hida, T.: White noise analysis and stochastic functional differential equations; *Studies in Modeling and Statistical Sci. Australian J. of Statistics, the J. Gani volume* **30A** (1988) 241–246

[153] Hida, T.: A note on generalized Gaussian random fields; *J. Multivariate Anal.* **27** (1988) 255–260

[154] Hida, T.: White noise and stochastic variational calculus for Gaussian random fields; in:*Dynamics and Stochastic Processes, Lec. Notes in Physics* **335** (1988) 136–141

[155] Hida, T.: Infinite-dimensional rotation group and unitary group; *Lecture Notes in Math* **1379** (1989) 125–134, Springer-Verlag

[156] Hida, T.: White noise analysis and Gaussian random fields; *Proc. the 24th Winter School of Theoretical Physics, Karpacz, Stochastic Methods in Math. and Phys.*, R. Gielerak and W. Karwowski (eds.) (1989) 277–289

[157] Hida, T.: Functionals of Brownian motion; *Lectures in Applied Mathematics and Informatics*, Luigi M. Ricciardi (ed.), Manchester Univ. Press, 1990, 286-329.

[158] Hida, T.: White noise and random fields - old and new; *Proc. Gaussian Random Fields*, T. Hida and K. Saitô (eds.) (1991) 1-10

[159] Hida, T.: Stochastic variational calculus; *Lecture Notes in Control and Information Sciences* **176** (1992) 123–134, Springer-Verlag

[160] Hida, T.: White noise and Gaussian random fields; in: *Probability Theory*, Louis H. Y. Chen (ed.), Walter de Ruyter & Co. (1992) 83–90

[161] Hida, T.: The impact of classical functional analysis on white noise calculus; *Centro Vito Volterra, Universita Degli Studi Di Roma II* (1992) #90

[162] Hida, T.: A role of the Lévy Laplacian in the causal calculus of generalized white noise functionals; in: *Stochastic Processes, G. Kallianpur Volume*, S. Cambanis (ed.) (1993) 131–139

[163] Hida, T.: White noise analysis and applications; *Stochastic Analysis and Applications in Physics*, A. I. Cardoso et al. (eds.) (1994) 119–131, Kluwer Academic Publishers

[164] Hida, T.: Some recent results in white noise analysis; *Pitman Research Notes in Math. Series* **310** (1994) 111–116, Longman Scientific & Technical

[165] Hida, T.: Random fields as generalized white noise functionals; *Acta Appl. Math.* **35** (1994) 49–61

[166] Hida, T.: Analysis of random functionals - Theory of nonlinear functions and applications (in Japanese); *Kagaku* **64** (1994) 334–339

[167] Hida, T.: White noise analysis: An overview and some future directions; *IIAS Reports* 1995-001 (1995)

[168] Hida, T.: White noise analysis and applications in random fields; in: *Proc. of the Conference on Dirichlet Forms and Stochastic Processes*, Z. Ma et al. (eds.) Walter de Gruiter (1995) 185–189

[169] Hida, T.: Infinite dimensional rotation group and white noise analysis; in: *Group Theoretical Methods in Physics, Proc. 20th Coloq. on Group Theoretical Methods in Physics*, A. Arima et al. (eds.) World Scientific (1995) 1–9

[170] Hida, T.: A note on stochastic variational equations; in: *Exploring Stochastic Laws. Korolyuk Volume*, A. V. Skorohod and Yu. V. Borovskikh (eds.) (1995) 147–152

[171] Hida, T.: Random fields and quantum dynamics; *Foundations of Physics* **27** (1997) 1511–1518

[172] Hida, T.: Fluctuation, nonlinearity and for human beings; *J. Tokyo University of Information Sciences* **2** (1998) 169–177

[173] Hida, T.: White noise approach to fluctuations; *J. Korean Math. Soc.* **35** (1998) 575-581

[174] Hida, T.: Some of future directions in white noise analysis; *Quantum Information*, T. Hida and K. Saitô (eds.) (1999) 103–110

[175] Hida, T.: White noise analysis and quantum dynamics; *Mathematical Methods of Quantum Physics, H. Ezawa Volume* (1999) 3–8

[176] Hida, T.: Harmonic analysis on complex random systems; in: *Infinite Dimensional Harmonic Analysis. Transactions of Japanese-Germany Symp.*, H. Heyer and T. Hirai (eds.) (1999) 160–166

[177] Hida, T.: Complexity in white noise analysis; *Quantum Information II*, Hida and K. Saitô (eds.) (2000) 61–70

[178] Hida, T.: White noise approach to Feynman integrals; in: *Proceedings of Feynman integral Conference*, Seoul, 1999 (to appear)

[179] Hida, T.: Complexity and irreversibility in stochastic analysis; in: *Proceeding of the Les Treilles Conference*, Solvay Inst., 1999 (to appear)

[180] Hida, T.: White noise theory and physics; in: *Proceeding of Millennial Symposium: Defining the Science Stochastics*, Wuerzburg. 2000 (to appear)

[181] Hida, T.: Innovation approach to random complex systems; *Pub. Centro Vito Volterra* **433** (2000)

[182] Hida, T.: Some methods of computation in white noise calculus; in: *Unconventional Model of Computation UMC'2K, Solvay Institute*, I. Antoniou et al. (eds.) (2001) 85–93

[183] Hida, T.: Innovation approach to stochastic process and quantum dynamics; *Proc. Vaxjo Conference on Foundations of Probability and Physics*, World Scientific Publ. (2001)

[184] Hida, T.: *White Noise and Functional Analysis* (to appear)

[185] Hida, T. and Kallianpur, G.: The square of a Gaussian Markov process and nonlinear prediction; *J. of Multivariate Analysis* **5** (1975) 451–461

[186] Hida, T., Kubo, I., Nomoto, H., and Yoshizawa, H.: On projective invariance of Brownian motion; *Publ. RIMS Kyoto Univ.*, **A 4** (1968) 595–609

[187] Hida, T. and Kuo, H.-H.: Semigroups associated with generalized Brownian functionals; *Semigroup Forum* **45** (1992) 261–263

[188] Hida, T., Kuo, H.-H., and Obata, N.: Transformations for white noise functionals; *J. Funct. Anal.* **111** (1993) 259–277

[189] Hida, T., Kuo, H.-H., Potthoff, J., and Streit, L.: *White Noise: An Infinite Dimensional Calculus*. Kluwer Academic Publishers, 1993

[190] Hida, T., Lee, K.-S., and Lee, S.-S.: Conformal invariance of white noise; *Nagoya Math. J.* **98** (1985) 87–98

[191] Hida, T., Lee, K.-S., and Si Si: Multidimensional parameter white noise and Gaussian random fields; *Balakrishnan Volume*, A. B. Aries (ed.) (1987) 177–183, Optimization Software

[192] Hida, T. and Mimachi, Y.: Some thoughts on infinite dimensional rotation group; *Quantum Information* **4**, World Scientific (2001)

[193] Hida, T., Obata, N., and Saitô: Infinite dimensional rotations and Laplacians in terms of white noise calculus; *Nagoya Math. J.* **128** (1992) 65–93

[194] Hida, T. and Potthoff, J.: White noise analysis - An overview; in: *White Noise Analysis, Mathematics and Applications* (1990) 140-165

[195] Hida, T., Potthoff, J., and Streit, L.: Dirichlet forms and white noise analysis; *Commun. Math. Phys.* **116** (1988) 235–245

[196] Hida, T., Potthoff, J., and Streit, L.: White noise analysis and applications; *Mathematics + Physics* **3** (1988) 143-178

[197] Hida, T., Potthoff, J., and Streit, L.: Energy forms and white noise analysis; *New Methods and Results in Nonlinear Field Equations, Lec, Notes in Phys.* **347** (1989) 115-125

[198] Hida, T. and Saitô, K.: White noise analysis and the Lévy Laplacian; *Stochastic Processes in Physics and Engineering*, S. Albeverio et al. (eds.) (1988) 177-184, Reidel Publishing Company

[199] Hida, T. and Saitô, K.: Introduction to white noise analysis; *Bielefeld Encounters in Math. and Phys. VII. BIBOS Publications* **400** (1989)

[200] Hida, T. and Si Si: Variational calculus for Gaussian random fields; *Proc. IFIP Warsaw, Lec. Notes in Control and Information Sci.* **136** (1989) 86-97

[201] Hida, T. and Si Si: Stochastic variational equations and innovations for random fields; in: *Infinite Dimensional Harmonic Analysis. Transactions of German-Japanese Symposium*, H. Heyer and T. Hirai (eds.) (1995) 86-93

[202] Hida, T. and Si Si: Innovation for random fields; *Infinite Dimensional Analysis, Quantum Probability and Related Topics* **1** (1998) 499-509

[203] Hida, T. and Si Si: Elemental random variables in white noise theory, beyond Reductionism; *Quantum Information* **3** (2001) (to appear)

[204] Hida, T. and Streit, L.: On quantum theory in terms of white noise; *Nagoya Math. J.* **68** (1977) 21-34

[205] Hida, T. and Streit, L.: Generalized Brownian functionals; *Proc. VI-th International Conference on Mathematical Physics, Lecture Notes in Physics* **153** (1982) 285-287

[206] Hitsuda, M.: Formula for Brownian partial derivatives; *Second Japan-USSR Symp. Probab. Th.* **2** (1972) 111-114

[207] Hitsuda, M.: Formula for Brownian partial derivatives; *Publ. Fac. Integrated Arts and Sciences, Hiroshima University, Series III* **4** (1978) 1-15

[208] Holden, H., Lindstrøm, T., Øksendal, B., and Ubøe, J.: Discrete Wick calculus and stochastic functional equations; *Potential Analysis* **1** (1992) 291-306

[209] Holden, H., Lindstrøm, T., Øksendal, B., and Ubøe, J.: Discrete Wick products; in: *Stochastic Analysis and Related Topics (4)*, T. Lindstrøm et al. (eds.), Gordon and Breach (1993) 123-148

[210] Holden, H., Lindstrøm, T., Øksendal, B., Ubøe, J., and Zhang, T.: Stochastic boundary value problems. A white noise functional approach; *Probability Theory and Related Fields* **95** (1993) 391-419

[211] Holden, H., Lindstrøm, T., Øksendal, B., Ubøe, J., and Zhang, T.: A comparison experiment for Wick multiplication and ordinary multiplication; in: *Stochastic Analysis and Related Topics (4)*, T. Lindstrøm et al. (eds.), Gordon and Breach (1993) 149-160

[212] Holden, H., Lindstrøm, T., Øksendal, B., Ubøe, J., and Zhang, T.: The Burgers equation with a noisy force and the stochastic heat equation; *Comm. PDE* **19** (1994) 119-141

[213] Holden, H., Lindstrøm, T., Øksendal, B., Ubøe, J., and Zhang, T.: The stochastic Wick-type Burgers equation; in: *Stochastic Partial Differential Equations*, A. Etheridge (ed.), Cambridge Univ. Press (1995) 141-161

[214] Holden, H., Lindstrøm, T., Øksendal, B., Ubøe, J., and Zhang, T.: The pressure equation for fluid flow in a stochastic medium; *Potential Analysis* **4** (1995) 655–674

[215] Holden, H. and Øksendal, B.: A white noise approach to stochastic differential equations driven by Wiener and Poisson noises; in: *Nonlinear Theory of Generalized Functions*, M. Grosser et al. (eds.) Chapman & Hall/CRC (1999) 293–313

[216] Holden, H. and Øksendal, B.: A white noise approach to stochastic Neumann boundary value problems; *Preprint University of Oslo* 8/1999 and *Acta Appl. Math.* (to appear)

[217] Holden, H., Øksendal, B., Ubøe, J., and Zhang, T.: *Stochastic Partial Differential Equations.* Birkhäuser, 1996

[218] Hu, Y., Lindstrøm, T., Øksendal, B., Ubøe, J., and Zhang, T.: Inverse powers of white noise; in: *Stochastic Analysis*, M. G. Cranston and M. Pinsky (eds.), American Mathematical Society (1995) 439–456

[219] Hu, Y. and Øksendal, B.: Wick approximation of quasilinear stochastic differential equations; in: *Stochastic Analysis and Related Topics (5)*, H. Korezlioglu (ed.) Birkhäuser (1996) 203–231

[220] Hu, Y. and Øksendal, B.: Fractional white noise calculus and applications to finance; *Preprint, University of Oslo* 10/1999

[221] Hu, Y. and Øksendal, B.: Chaos expansion of local time of fractional Brownian motions; *Preprint, University of Oslo* 20/2000

[222] Hu, Y., Øksendal, B., and Salopek, D. M.: Weighted local time for fractional Brownian motion and applications to finance; *Preprint* (2000)

[223] Hu, Y., Øksendal, B., and Sulem, A.: Optimal consumption and portfolio in a Black-Scholes market driven by fractional Brownian motion; *Preprint, University of Oslo* 23/2000

[224] Hu, Y. and Øksendal, B., and Sulem, A.: A stochastic maximum principle for processes driven by fractional Brownian motion; *Preprint, University of Oslo* 24/2000

[225] Hu, Y. and Øksendal, B., and Zhang, T.: Stochastic partial differential equations driven by multiparameter fractional white noise; in: *Stochastic Processes, Physics and Geometry: New Interplays II*, a volume in honor of S. Albeverio, F. Gesztesy et al. (eds.) American Mathematical Society (2000) 327–337

[226] Huang, Z.-Y.: Stochastic calculus of variation on Gaussian spaces and white noise analysis; in: *Gaussian random fields*, World Scientific (1991) 227–241

[227] Huang, Z.-Y.: Quantum white noises–White noise approach to quantum stochastic calculus; *Nagoya Math. J.* **129** (1993) 23–42

[228] Ito, Y.: On a generalization of nonlinear Poisson functionals; *Math. Rep. Toyama Univ.* **3** (1980) 111–122

[229] Ito, Y.: Generalized Poisson functionals; *Probab. Theory Related Fields* **77** (1988) 1–28

[230] Ito, Y. and Kubo, I.: Calculus on Gaussian and Poisson white noises; *Nagoya Math. J.* **111** (1988) 41–84

[231] Ito, Y., Kubo, I. and Takenaka, S.: Calculus on Gaussian white noise and Kuo's Fourier transformation; *White Noise Analysis–Mathematics and Applications*, T. Hida et al. (eds.) (1990) 180-207, World Scientific

[232] Ji, U. C. and Obata, N.: Initial value problem for white noise operators and quantum stochastic processes; in: *Infinite Dimensional Harmonic Analysis*, H. Heyer et al. (eds.), D.+M. Gräbner (2000) 203–216

[233] Kallianpur, G. and Kuo, H.-H.: Regularity property of Donsker's delta function; *Appl. Math. Optim.* **12** (1984) 89–95

[234] Kang, S.-J.: Heat and Poisson equations associated with number operator in white noise analysis; *Soochow J. Math.* **20** (1994) 45–55

[235] Karwowski, W. and Streit, L.: A renormalization group model with non-Gaussian fixed point; *Rep. Math. Phys* **13** (1978) 1

[236] Khandekar, D. C. and Streit, L.: Constructing the Feynman integrand; *Annalen der Physik* **1** (1992) 49–55

[237] Khrennikov, A. and Huang, Z.: A model for white noise analysis in p-adic number fields; *Acta Math. Sci.* **16** (1996) 1–14

[238] Kondratiev, Yu. G., Leukert, P., Potthoff, J., Streit, L., and Westerkamp, W.: Generalized functionals in Gaussian spaces–the characterization theorem revisited; *J. Funct. Analysis* **141** (1996) 301–318

[239] Kondratiev, Yu. G., Leukert, P., and Streit, L.: Wick calculus in Gaussian analysis; *Acta Appl. Math* **44** (1996) 269–294

[240] Kondratiev, Yu. G., da Silva, J. L., and Streit, L.: Generalized Appell systems; *Meth. Funct. Anal. Top.* **3** (1997) 28–61

[241] Kondratiev, Yu. G., Silva, J. L., and Streit, L.: Differential geometry on compound Poisson space; *Methods Funct. Anal. Topology* **4** (1998) 32–58

[242] Kondratiev, Yu. G., da Silva, J. L., Streit, L., and Us, G.: Analysis on Poisson and gamma spaces; *Inf. Dim. Anal. Qu. Prob.* **1** (1998) 91–117

[243] Kondratiev, Yu. G. and Streit, L.: A remark about a norm estimate for white noise distributions; *Ukrainean Math. J.* **44** (1992) 832–835

[244] Kondratiev, Yu. G. and Streit, L.: Spaces of white noise distributions: Constructions, Descriptions, Applications. I; *Reports on Math. Phys.* **33** (1993) 341–366

[245] Kondratiev, Yu. G., Streit, L., and Westerkamp, W.: A note on positive distributions in Gaussian analysis; *Ukrainian Math. J.* **47** (1995) 749–759

[246] Kondratiev, Yu. G., Streit, L., and Westerkamp, W., and Yan, J. A.: Generalized functions in infinite dimensional analysis; *Hiroshima Math. Journal* **28** (1998) 213–260

[247] Kubo, I.: Itô formula for generalized Brownian functionals; *Lecture Notes in Control and Information Scis.* **49** (1983) 156–166, Springer-Verlag

[248] Kubo, I.: The structure of Hida distributions; in: *Mathematical Approach to Fluctuations I*, World Scientific (1994) 49–114

[249] Kubo, I.: A direct setting of white noise calculus; *Pitman Research Notes in Math. Series* **310** (1994) 152–166, Longman Scientific & Technical

[250] Kubo, I.: Weighted number operators and associated process; in: *Infinite Dimensional Harmonic Analysis* (1996) 146–155

[251] Kubo, I.: Generalized functionals in white noise analysis; in: *Probability Theory and Mathematical Statistics*, Word Scientific (1996) 237–243

[252] Kubo, I.: Non-isotropic Ornstein-Uhlenbeck process and white noise analysis; in: *Stochastic Differential and Difference Equations*, Birkhäuser (1997) 167–182

[253] Kubo, I.: Entire Functionals and generalized functionals in white noise analysis; in: *Analysis on infinite-dimensional Lie groups and algebra*, World Scientific (1998) 207–215

[254] Kubo, I. and Kuo, H.-H.: Fourier transform and cylindrical Hida distributions; in *Stochastic Processes, a Festschrift in Honor of G. Kallianpur*, S. Cambanis et al. (eds.), Springer-Verlag (1993) 191–201

[255] Kubo, I. and Kuo, H.-H.: Finite dimensional Hida distributions; *J. Funct. Anal.* **128** (1995) 1–47

[256] Kubo, I. and Kuo, H.-H.: A simple proof of Hida distribution characterization theorem; *Exploring Stochastic Laws, Festschrift in Honor of the 70th Birthday of V. S. Korolyuk*, A. V. Skorokhod & Yu. V. Borovskikh (eds.) (1995) 243–250, International Science Publishers

[257] Kubo, I., Kuo, H.-H., and Sengupta, A.: White noise analysis on a new space of Hida distributions; *Infinite Dimensional Analysis, Quantum Probability, and Related Topics* **2** (1999) 315–335

[258] Kubo, I. and Lee, S.-S.: Non-linear predictors of transformed stationary processes; *Nagoya Math. J.* **95** (1984) 23–40

[259] Kubo, I. and Mustafid: Limit theorem of symmetric statistics represented by multiple Poisson-Wiener integrals; in: *Abstract of Papers of Supplementary meeting to the 15th SPA Conference* (1986) 17–19

[260] Kubo, I. and Takenaka, S.: Calculus on Gaussian white noise I; *Proc. Japan Acad.* **56A** (1980) 376–380

[261] Kubo, I. and Takenaka, S.: Calculus on Gaussian white noise II; *Proc. Japan Acad.* **56A** (1980) 411–416

[262] Kubo, I. and Takenaka, S.: Calculus on Gaussian white noise III; *Proc. Japan Acad.* **57A** (1981) 433–437

[263] Kubo, I. and Takenaka, S.: Calculus on Gaussian white noise IV; *Proc. Japan Acad.* **58A** (1982) 186–189

[264] Kubo, I., Takenaka, S., and Urakawa, H.: Brownian motion parametrized with metric space of constant curvature; *Nagoya Math. J.* **82** (1981) 131–140

[265] Kubo, I. and Yokoi, Y.: A remark on the space of testing random variables in the white noise calculus; *Nagoya Math. J.* **115** (1989) 139–149

[266] Kubo, I. and Yokoi, Y.: Generalized functions and functionals in fluctuation analysis; in: *Mathematical Approach to Fluctuations II*, World Scientific (1995) 203–230

[267] Kuna, T., Streit, L., and Westerkamp, W.: Feynman integrals for a class of exponentially growing potentials; *J. Math. Phys.* **39** (1998) 4476–4491

[268] Kuo, H.-H.: Fourier-Wiener transform on Brownian functionals; *Lecture Notes in Math., Springer-Verlag* **828** (1980) 146–161

[269] Kuo, H.-H.: On Fourier transform of generalized Brownian functionals; *J. Multivariate Analysis* **12** (1982) 415–431

[270] Kuo, H.-H.: Donsker's delta function as a generalized Brownian functional and its application; *Lecture Notes in Control and Information Sciences* **49** (1983) 167–178, Springer-Verlag

[271] Kuo, H.-H.: Fourier-Mehler transforms of generalized Brownian functionals; *Proc. Japan Acad.* **59A** (1983) 312–314

[272] Kuo, H.-H.: Brownian functionals and applications; *Acta Appl. Math.* **1** (1983) 175–188

[273] Kuo, H.-H.: A Fourier transform characterization of Gaussian Brownian functionals; *Bull. Inst. Math. Academia Sinica* **11** (1983) 407–413

[274] Kuo, H.-H.: Brownian functionals and applications; *Acta Applicandae Mathematicae* **1** (1983) 175–188

[275] Kuo, H.-H.: Fourier-Mehler transforms of generalized Brownian functionals; *Proc. Japan Academy* **59A** (1983) 312–314

[276] Kuo, H.-H.: On Laplacian operators of generalized Brownian functionals; *Lecture Notes in Math.* **1203** (1986) 119–128, Springer-Verlag

[277] Kuo, H.-H.: The heat equation and the Fourier transform of generalized Brownian functionals; *Lecture Notes in Math., Springer-Verlag* **1236** (1987) 154–163

[278] Kuo, H.-H.: Brownian motion, diffusions and infinite dimensional calculus; *Lecture Notes in Math., Springer-Verlag* **1316** (1988) 130–169

[279] Kuo, H.-H.: White noise calculus; in *Algebra, Analysis and Geometry*, M. C. Kang and K. W. Lih (eds.), World Scientific (1989) 77–119

[280] Kuo, H.-H.: Stochastic partial differential equations of generalized Brownian functionals; *Lecture Notes in Math., Springer-Verlag* **1390** (1989) 138–146

[281] Kuo, H.-H.: The Fourier transform in white noise calculus; *J. Multivariate Analysis* **31** (1989) 311-327

[282] Kuo, H.-H.: Fourier-Mehler transforms in white noise analysis; *Gaussian Random Fields*, K. Itô and T. Hida (eds.) (1991) 257–271, World Scientific

[283] Kuo, H.-H.: Lectures on white noise analysis; *Soochow J. Math.* **18** (1992) 229–300

[284] Kuo, H.-H.: Convolution and Fourier transform of Hida distributions; *Lecture Notes in Control and Information Sciences* **176** (1992) 165–176, Springer-Verlag

[285] Kuo, H.-H.: An introduction to white noise calculus; *Aportaciones Matemáticas, Notas de Investigación* **7** (1992) 1–12

[286] Kuo, H.-H.: *Infinite Dimensional Stochastic Analysis*. Nagoya University Lecture Notes in Math. **11** (1993)

[287] Kuo, H.-H.: Analysis of white noise functionals; *Soochow J. Math.* **20** (1994) 419–464

[288] Kuo, H.-H.: White noise analysis; in *Proc. Workshops in Pure Math.* **14**, part II: Analysis, D. Kim (ed.) (1994) 27–40, Pure Math. Research Assoc., Korean Academic Council

[289] Kuo, H.-H.: An infinite dimensional Fourier transform; in *Stochastic Processes, Physics, and Geometry II*, S. Albeverio et al. (eds.), World Scientific (1995) 482–490

[290] Kuo, H.-H.: *White Noise Distribution Theory*. CRC Press, 1996

[291] Kuo, H.-H.: A characterization of Hida measures; in *Stochastic Processes and Functional Analysis*, J. A. Goldstein et al. (eds.), Marcel Dekker (1997) 147–151

[292] Kuo, H.-H.: Stochastic integration via white noise analysis; *Nonlinear Analysis, Theory, Methods, and Applications* **30** (1997) 317–328

[293] Kuo, H.-H.: Anticipatory Itô's formula and Hitsuda-Skorokhod integral; in *Skorokhod's Ideas in Probability Theory*, V. Korolyuk et al. (eds.), Proc. Institute of Math., National Academy of Sciences of Ukraine, Vol. 32 (2000) 248–255

[294] Kuo, H.-H.: Growth functions in white noise theory; *Soochow J. Math.* **26** (2000) 327–342

[295] Kuo, H.-H.: Some aspects of white noise analysis; in *Proceedings of the Volterra International School on White Noise Approach to Classical and Quantum Stochastic Calculi*, L. Accardi (ed.) World Scientific (to appear)

[296] Kuo, H.-H.: White noise theory; in *Handbook of Stochastic Analysis and Applications*, D. Kannan and V. Lakshmikantham (eds.) Marcel Dekker (to appear)

[297] Kuo, H.-H., Lee, Y.-J., and Shih, C.-Y.: Wiener-Itô theorem in terms of Wick tensors; *Acta Applicandae Mathematicae* **63** (2000) 203–218

[298] Kuo, H.-H. and Nishi, K.: An anticipatory Itô formula; *Proc. Japan Academy* **72A** (1996) 36–38

[299] Kuo, H.-H., Obata, N., and Saitô, K.: Lévy Laplacian of generalized functionals on a nuclear space; *J. Funct. Anal.* **94** (1990) 74–92

[300] Kuo, H.-H., Obata, N., and Saitô, K.: Diagonalization of the Lévy Laplacian and related stable processes; *Preprint* (2001)

[301] Kuo, H.-H., Obata, N., and Saitô, K.: White noise analysis based on the Leévy Laplacian; *Preprint* (2001)

[302] Kuo, H.-H. and Potthoff, J.: Anticipating stochastic integrals and stochastic differential equations; *White Noise Analysis–Mathematics and Applications*, T. Hida et al. (eds.) (1990) 256–273, World Scientific

[303] Kuo, H.-H. and Potthoff, J.: Anticipating stochastic differential equations; in *Probability Theory and Math. Stat., Proc. 5th Intern. Conference*, vol. **2**, 35-46, Vilnius, Lithuania (1990)

[304] Kuo, H.-H., Potthoff, J., and Streit, L.: A characterization of white noise test functionals; *Nagoya Math. J.* **121** (1991) 185–194

[305] Kuo, H.-H., Potthoff, J., and Yan, J.-A.: Continuity of affine transformations of white noise test functionals and applications; *Stochastic Processes and Their Applications* **43** (1992) 85–98

[306] Kuo, H.-H. and Russek, A.: White noise approach to stochastic integration; *J. Multivariate Analysis* **24** (1988) 218–236

[307] Kuo, H.-H., Saitô, K., and Stan, A.: A Hausdorff-Young inequality for white noise analysis; *Preprint* (2001)

[308] Kuo, H.-H. and Shieh, N. R.: A generalized Itô's formula for multidimensional Brownian motions and its applications; *Chinese J. Math.* **15** (1987) 163–174

[309] Kuo, H.-H. and Smolenski, W.: On admissible shifts of generalized white noises; *J. Multivariate Analysis* **12** (1982) 80–88

[310] Kuo, H.-H. and Sun, T.-C.: Absolute value of white noise as a generalized Brownian functional; *Soochow J. Math* **15** (1989) 205–211

[311] Kuo, H.-H. and Xiong, J.: Stochastic differential equations in white noise space; *Infinite Dimensional Analysis, Quantum Probability, and Related Topics* **1** (1998) 611–632

[312] Lascheck, A., Leukert, P., Streit, L., and Westerkamp, W.: Quantum mechanical propagators in terms of Hida distributions; *Rep. Math. Phys.* **33** (1993) 221–232

[313] Lascheck, A., Leukert, P., Streit, L., and Westerkamp, W.: More about Donsker's delta function; *Soochow J. Math.* **20** (1994) 401–418

[314] Lee, J.-S.: Properties of the Fourier transform in white noise analysis; *Kyung-pook Math. J.* **35** (1996) 695–704

[315] Lee, K. Sim: Gel'fand triples associated with finite-dimensional Gaussian measure; *Soochow J. Math.* **22** (1996) 1–16

[316] Lee, Y.-J.: Applications of the Fourier-Wiener transform to differential equations on infinite dimensional spaces, I; *Trans. Amer. Math. Soc.* **262** (1980) 259–283

[317] Lee, Y.-J.: Parabolic equations on infinite dimensional spaces; *Bull. Inst. Math. Acad. Sinica* **9** (1981) 279–292

[318] Lee, Y.-J.: Applications of the Fourier-Wiener transform to differential equations on infinite dimensional spaces, II; *J. Diff. Equa.* **41** (1981) 59–86

[319] Lee, Y.-J.: Fundamental solutions for differential equations associated with the number operator; *Trans. Amer. Soc.* **268** (1981) 467–476

[320] Lee, Y.-J.: Integral transforms of analytic functions on abstract Wiener spaces; *J. Funct. Anal.* **47** (1982) 153–164

[321] Lee, Y.-J.: A correction and some additions to the paper "TAMS 268 (1981) 467–476"; *Trans. Amer. Math. Soc.* **276** (1983) 621–623

[322] Lee, Y.-J.: Wiener process and Hardy spaces on abstract Wiener spaces; *Proc. NSC., Part A* **10** (1986) 275–280

[323] Lee, Y.-J.: Sharp inequalities and regularity of heat semigroup on infinite dimensional spaces; *J. Funct. Anal.* **71** (1987) 69–87

[324] Lee, Y.-J.: Unitary operators on the space of L^2–functions over abstract Wiener spaces; *Soochow J. Math.* **13** (1987) 165–174

[325] Lee, Y.-J.: Generalized functions on infinite dimensional spaces and its application to white noise calculus; *J. Funct. Anal.* **82** (1989) 429–464

[326] Lee, Y.-J.: On the convergence of Wiener–Ito decomposition; *Bull. Inst. Math. Acad. Sinica* **17** (1989) 305–312

[327] Lee, Y.-J.: A reformulation of white noise calculus; in: *White Noise Analysis – Mathematics and Applications*, T. Hida et al. (eds.) World Scientific (1990) 274–293

[328] Lee, Y.-J.: Analytic version of test functionals, Fourier transform and a characterization of measures in white noise calculus; *J. Funct. Anal.* **100** (1991) 359–380

[329] Lee, Y.-J.: A characterization of generalized functions on infinite-dimensional spaces and Bargman-Segal analytic functions; *Gaussian Random Fields*, K. Itô and T. Hida (eds.) (1991) 272–284, World Scientific

[330] Lee, Y.-J.: Positive generalized functions on infinite dimensional spaces; in: *Stochastic Processes, A Festschrift in Honour of Gopinath Kallianpur*, Springer-Verlag (1993) 225–234

[331] Lee, Y.-J.: Transformation and Wiener-Ito decomposition of white noise functionals; *Bull. Inst. Math. Acad. Sinica* **21** (1993) 279–291

[332] Lee, Y.-J.: Exact formula of certain functional integrals on Wiener spaces; *Stochastic and Stochastic Report* **50** (1994) 211–223

[333] Lee, Y.-J.: Transformation of white noise functionals and applications; *Pitman Research Notes in Mathematics Series* **310** (1994) 183–196

[334] Lee, Y.-J.: Integral representation of second quantization and its application in white noise analysis; *J. Funct. Anal.* **133** (1995) 253–276

[335] Lee, Y.-J.: A generalization of Mehler transform; in: *Proc. International Mathematics Conference*, Y. Fong et al. (eds.), World Scientific (1996)

[336] Lee, Y.-J.: Convergence of Fock expansion and transformations of Brownian functionals; in: *Functional Analysis and Global Analysis*, T. Sunada and P. W. Sy (eds.) (1996) 142-156

[337] Lee, Y.-J.: A generalization of Riesz representation theorem to infinite dimensions; *J. Funct. Anal.* **151** (1997) 121-137

[338] Lee, Y.-J.: Generalized white noise functionals on classical Wiener spaces; *J. Korean Math. Soc.* **35** (1998) 613-635

[339] Lee, Y.-J. and Shih, C.-Y.: A probabilistic approach to the integral representation of positive generalized white noise functionals; *Stochastic Analysis and its Applications* (to appear)

[340] Lee, Y.-J. and Shih, C.-Y.: The Riesz representation theorem in infinite dimensions and Its applications; *Infinite Dimensional Analysis, Quantum Probability and Related Topics* (to appear)

[341] Lee, Y.-J. and Shih, H.-H.: The Segal-Bargmann transform for Lévy functionals; *J. Funct. Anal.* **168** (1999) 46-83

[342] Lee, Y.-J. and Shih, H.-H.: Donsker-Delta function of Lévy process; *Acta Applicandae Mathematicae* **63** (2000) 219-231

[343] Lee, Y.-J. and Shih, H.-H.: Ito formula for generalized Lévy functionals; *Quantum Information II*, T. Hida and K. Saito (eds.), Word Scientific (2000) 87-105

[344] Lee, Y.-J. and Stan, A.: An infinite dimensional Heisenberg uncertainty principle; *Taiwanese J. Math.* **3** (1999) 529-538

[345] Lindstrøm, T., Øksendal, B., and Ubøe, J.: Stochastic differential equations involving positive noise; *Stochastic Analysis*, M. Barlow and N. Bingham (eds.) (1991) 261-303, Cambridge University Press

[346] Lindstrøm, T., Øksendal, B., and Ubøe, J.: Stochastic modelling of fluid flow in porous media; in: *Control Theory, Stochastic Analysis and Applications*, S. Chen and J. Yong (eds.), World Scientific (1991) 156-172

[347] Lindstrøm, T., Øksendal, B., and Ubøe, J.: Wick multiplication and Itô-Skorohod stochastic differential equations; in: *Ideas and Methods in Mathematical Analysis, Stochastics and Applications*, S. Albeverio et al. (eds.), Cambridge Univ. Press (1992) 183-206

[348] Lindstrøm, T., Øksendal, B., Ubøe, J., and Zhang, T.: Stability properties of stochastic partial differential equations; *Stochastic Analysis and Applications* **13** (1995) 177-204

[349] Luo, S. and Yan, J. A.: Generalized Fourier-Mehler transforms on white noise functional spaces; *Chinese Sci. Bull.* **43** (1998) 1321-1325

[350] Luo, S. and Yan, J. A.: Gaussian kernel operators on white noise functional spaces; *Sci. China Ser. A* **43** (2000) 1067-1074

[351] Meyer, P.-A. and Yan, J. A.: Distributions sur l'espace de Wiener (suite) d'apres I. Kubo et Y. Yokoi; *Lecture Notes in Math.* **1372** (1989) 382-392

[352] Ngobi, S. and Stan, A.: An extension of the Clark-Ocone formula; *Preprint* (2001)

[353] Nishi, K. and Saitô, K.: The Lévy Laplacian and the number operator; *RIMS Kokyuroku* **1193** (2001)

[354] Nishi, K., Saitô, K., and Tsoi, A. H.: A stochastic expression of a semigroup generated by the Lévy Laplacian; in: *Quantum Information III*, World Scientific (2001)

[355] Obata, N.: Certain unitary representations of the infinite symmetric group I; *Nagoya Math. J.* **105** (1987) 109–119

[356] Obata, N.: Certain unitary representations of the infinite symmetric group II; *Nagoya Math. J.* **106** (1987) 143–162

[357] Obata, N.: Analysis of the Lévy Laplacian; *Soochow J. Math.* **14** (1988) 105–109

[358] Obata, N.: A note on certain permutation groups in the infinite dimensional rotation group; *Nagoya Math. J.* **109** (1988) 91–107

[359] Obata, N.: Density of natural numbers and the Lévy group; *J. Number Theory* **30** (1988) 288–297

[360] Obata, N.: The Lévy Laplacian and the mean value theorem; *Lecture Notes in Math.* **1379** (1989) 242–253, Springer-Verlag

[361] Obata, N.: Some remarks on induced representations of infinite discrete groups; *Math. Ann.* **284** (1989) 91–102

[362] Obata, N.: A characterization of the Lévy Laplacian in terms of infinite dimensional rotation groups; *Nagoya Math. J.* **118** (1990) 111–132

[363] Obata, N.: Integral expression of some indecomposable characters of the infinite symmetric group in terms of irreducible representations; *Math. Ann.* **287** (1990) 369–375

[364] Obata, N.: Construction of irreducible unitary representations of amalgams of discrete abelian groups; *J. Math. Soc. Japan* **42** (1990) 585–603

[365] Obata, N.: Isometric operators on L^1-algebras of hypergroups; in: *Probability Measures on Groups X (H. Heyer Ed.)*, Plenum Press, New York (1991) 315–328

[366] Obata, N.: Rotation-invariant operators on white noise functionals; *Math. Z.* **210** (1992) 69–89

[367] Obata, N.: Elements of white noise calculus; *Centro V. Volterra publications* **124** (1992)

[368] Obata, N.: An analytic characterization of symbols of operators on white noise functionals; *J. Math. Soc. Japan* **45** (1993) 421–445

[369] Obata, N.: White noise delta functions and continuous version theorem; *Nagoya Math. J.* **129** (1993) 1–22

[370] Obata, N.: Operator calculus on vector-valued white noise functionals; *J. Funct. Anal.* **121** (1994) 185–232

[371] Obata, N.: *White Noise Calculus and Fock Space.* Lecture Notes in Math. **1577**, Springer-Verlag, 1994

[372] Obata, N.: Harmonic analysis and infinite dimensional Laplacians on Gaussian space; in: *Quantum Probability and Related Topics IX*, L. Accardi et al. (eds.), World Scientific (1994) 327–370

[373] Obata, N.: Conditional expectation in classical and quantum white noise calculi; *Analysis of Operators on Gaussian Space and Quantum Probability Theory*, RIMS Kokyuroku **923** (1995) 154–190

[374] Obata, N.: Fock expansion of operators on white noise functionals; in: *Stochastic Processes, Physics and Geometry II*, S. Albeverio et al. (eds.), World Scientific (1995) 557–568

[375] Obata, N.: Derivations on white noise functionals; *Nagoya Math. J.* **139** (1995) 21–36

[376] Obata, N.: Lie algebras containing infinite dimensional Laplacians; in: *Probability Measures on Groups and Related Structures*, H. Heyer (ed.), World Scientific (1995) 260–273

[377] Obata, N.: Generalized quantum stochastic processes on Fock space; *Publ. RIMS, Kyoto Univ.* **31** (1995) 667–702

[378] Obata, N.: White noise approach to quantum martingales; in: *Probability Theory and Mathematical Statistics*, M. Fukushima et al. (eds.), World Scientific (1996) 379–386

[379] Obata, N.: Invitation to quantum white noise; in: *Dynamical Systems in Infinite Dimensional Spaces*, Y. M. Park (ed.), Pure Math. Res. Assoc., Korean Academic Council (1996) 1–34

[380] Obata, N.: Constructing one-parameter transformation groups on white noise functions in terms of equicontinuous generators; in: *Infinite Dimensional Harmonic Analysis*, H. Heyer and T. Hirai (eds.), Gräbner (1996) 169–187

[381] Obata, N.: Integral kernel operators on Fock space – Generalizations and applications to quantum dynamics; *Acta Appl. Math.* **47** (1997) 49–77

[382] Obata, N.: Quantum stochastic differential equations in terms of quantum white noise; *Nonlinear Analysis, Theory, Methods and Applications* **30** (1997) 279–290

[383] Obata, N.: Time-ordered Wick exponential and quantum stochastic differential equations; in: *Quantum Communication, Computing, and Measurement*, C. M. Caves et al. (eds.), Plenum (1997) 355–363

[384] Obata, N.: Constructing one-parameter transformation groups on white noise functions in terms of equicontinuous generators; *Monatsh. Math.* **124** (1997) 317–335

[385] Obata, N.: A note on Hida's whiskers and complex white noise; in: *Analysis on Infinite-Dimensional Lie Groups and Algebras*, H. Heyer and J. Marion (eds.), World Scientific (1998) 321–336

[386] Obata, N.: A note on normal-ordered white noise equations; in: *Invited Lectures at the VIII-th Internat. Colloq. Differential Equations* II, D. Kolev (ed.), Academic Publications (1999) 109–118

[387] Obata, N.: Wick product of white noise operators and quantum stochastic differential equations; *J. Math. Soc. Japan* **51** (1999) 613–641

[388] Obata, N.: A note on coherent state representations of white noise operators; in: *Quantum Information II*, T. Hida and K. Saitô (eds.), World Scientific (2000) 135–147

[389] Obata, N.: Quantum stochastic analysis via white noise operator theory; in: *Proc. Internat. Conf. on Probability and Statistics and Their Applications*, Hanoi Institute of Mathematics (2000) 41–65

[390] Obata, N.: Coherent state representations in white noise calculus; *Can. Math. Soc. Conference Proceedings* **29** (2000) 517–531

[391] Obata, N.: White noise operators associated with the Cochran–Kuo–Sengupta space; *Proc. Japan–Germany Seminar* (2000)

[392] Obata, N.: Complex white noise and coherent state representation; *Acta Applicandae Mathematicae* **63** (2000)

[393] Obata, N.: White noise operator theory: Fundamental concepts and developing applications; *Proceedings of Volterra International School* (2000)

[394] Obata, N.: Coherent state representation and unitarity condition in white noise calculus; *J. Korean Math. Soc.* **38** (2001)

[395] Obata, N.: Unitarity criterion in white noise calculus and nonexistence of unitary evolutions driven by higher powers of quantum white noises; *Proc. Mexico Guanajuato Conference* (2001)

[396] Obata, N.: Quadratic quantum white noises and Lévy Laplacian; *Nonlinear Analysis* (2001)

[397] Obata, N. and Wildberger, N. J.: Generalized hypergroups and orthogonal polynomials; *Nagoya Math. J.* **142** (1996) 67–93

[398] Øksendal, B.: Stochastic partial differential equations: A mathematical connection between macrocosmos and microcosmos; in: *Analysis, Algebra, and Computers in Mathematical Research*, M. Gyllenberg and L. E. Persson (eds.), Marcel Dekker (1994) 365–385

[399] Øksendal, B.: Some mathematical models for population growth in a stochastic environment; in: *Proc. 9th SAMSA Symposium "Mathematics and the Environment"*, R. K. Colegrave et al. (eds.), University of Botswana (1994) 40–54

[400] Øksendal, B.: Stochastic partial differential equations and applications to hydrodynamics; in: *Stochastic Analysis and Applications in Physics*, A. I. Cardoso et al. (eds.), NATO ASI Series, Vol.449. Kluwer (1994) 283–305

[401] Øksendal, B.: *An Introduction to Malliavin Calculus with Applications to Economics*. Lecture Notes from a course given 1996 at the Norwegian School of Economics and Business Administration (NHH), NHH Preprint Series, 1996

[402] Øksendal, B. and Vøge, G.: A moving boundary problem in a stochastic medium; *Infinite Dimensional Analysis, Quantum Probability and Related Topics* **2** (1999) 179–202

[403] Øksendal, B. and Zhang, T.: The stochastic Volterra equation; in: *Barcelona Seminar on Stochastic Analysis*, D. Nualart and M. Sanz Sole (eds.), Birkhäuser (1993) 168–202

[404] Øksendal, B. and Zhang, T.: The general linear stochastic Volterra equation with anticipating coefficients; in: *Stochastic Analysis and Applications*, I. M. Davis et al. (eds.), World Scientific (1996) 343–366

[405] Øksendal, B. and Zhang, T.: Multiparameter fractional Brownian motion and quasi-linear stochastic partial differential equations; *Preprint, University of Oslo* 5/2000 and *Stochastics and Stochastics Reports* (to appear)

[406] Ouerdiane, H. and Rezgui, A.: Un théorème de Bochner-Minlos avec une condition d'intégrabilité; *Infinite Dimensional Analysis, Quantum Probability and Related Topics* **3** (2000) 297–302

[407] Potthoff, J.: On the connection of the white-noise and Malliavin calculi; *Proc. Japan Acad.* **62** (1986) 43–45

[408] Potthoff, J.: White-noise approach to Malliavin's calculus; *J. Funct. Anal.* **71** (1987) 207–217

[409] Potthoff, J.: On positive generalized functionals; *J. Funct. Anal.* **74** (1987) 81–95

[410] Potthoff, J.: Littlewood - Paley theory on Gaussian spaces; *Nagoya Math. J.* **109** (1988) 47–61

[411] Potthoff, J.: On Meyer's equivalence; *Nagoya Math. J.* **111** (1988) 99–109

[412] Potthoff, J.: On the construction of Dirichlet forms in infinite dimensions; in: *Proc. of the IX-th Int. Congr. Math. Phys.* B. Simon et al. (eds.), Adam Hilger (1989)

[413] Potthoff, J.: Stochastic integration in Hida's white noise analysis; in: *Stochastic Processes, Physics and Geometry*, S. Albeverio et al. (eds.), World Scientific (1990)

[414] Potthoff, J.: Introduction to white noise analysis; in: *Control Theory, Stochastic Analysis and Applications*, S. Chen and J. Yong (eds.), World Scientific (1991)

[415] Potthoff, J.: White noise methods for stochastic partial differential equations; in: *Stochastic Partial Differential Equations and Their Applications*, B.L. Rozovskii and R.B. Sowers (eds.), Springer (1992)

[416] Potthoff, J.: White noise approach to parabolic stochastic partial differential equations; in: *Stochastic Analysis and Applications in Physics*, A.I. Cardoso et al. (eds.), Kluwer Academic Publishers (1994)

[417] Potthoff, J.: On differential operators in white noise analysis; *Acta Appl. Math.*, a special volume in honor of T. Hida (2001)

[418] Potthoff, J. and Röckner, M.: On the contraction property of energy forms in infinite dimensions; *J. Funct. Anal.* **92** (1990) 155–165

[419] Potthoff, J. and Smajlovic, E.: On Donsker's delta function in white noise analysis; in: *Mathematical Physics and Stochastic Analysis* - Essays in honour of L. Streit, S. Albeverio et al. (eds.), World Scientific (2000)

[420] Potthoff, J. and Streit, L.: A characterization of Hida distributions; *J. Funct. Anal.* **101** (1991) 212–229

[421] Potthoff, J. and Streit, L.: Generalized Radon-Nikodym derivatives and Cameron-Martin theory; *Gaussian Random Fields*, K. Itô and T. Hida (eds.) (1991) 320–331, World Scientific

[422] Potthoff, J. and Streit, L.: White noise analysis and what it can do for physics; in: *Gaussian Random Fields*, K. Itô and T. Hida (eds.), World Scientific (1991) 58–68

[423] Potthoff, J. and Streit, L.: Invariant states on random and quantum fields: ϕ bounds and white noise analysis; *J. Funct. Anal.* **111** (1993) 295–311

[424] Potthoff, J. and Sundar, P.: Law of large numbers and central limit theorem for Donsker's Delta function; *Stochastics and Stochastic Reports* **43** (1993) 135–150

[425] Potthoff, J. and Sundar, P.: Limit theorems for the Donsker delta function: An example; in: *Stochastic Analysis and Related Topics*, T. Lindstrøm (eds.), Gordon and Breach (1993)

[426] Potthoff, J. and Sundar, P.: Law of large numbers and central limit theorem for Donsker's delta function of diffusions I; *Potential Analysis* **5** (1996) 487–504

[427] Potthoff, J. and Timpel, M.: On a dual pair of spaces of random variables; *Potential Analysis* **4** (1995) 637–654

[428] Potthoff, J., Vage, G., and Watanabe, H.: Generalized solutions of linear parabolic stochastic partial differential equations; *Appl. Math. Optimization* **38** (1998) 95–107

[429] Potthoff, J. and Yan, J. A.: Some results about test and generalized functionals of white noise; *Probability Theory*, L. H. Y. Chen et al. (eds.) (1992) 121–145, Walter de Gruyter & Co.

[430] Redfern, M.: White noise approach to multiparameter stochastic integration; *J. Multivariate Analysis* **37** (1991) 1–23

[431] Redfern, M.: Stochastic integration via white noise and the fundamental theorem of calculus; *Stochastic analysis on infinite dimensional spaces*, H. Kunita and H.-H. Kuo (eds.), Longman House, (1994) 255–263

[432] Redfern, M.: Two-parameter Stratonovich integrals; *Dynam. Contin. Discrete Impuls. Systems* **5** (1999) 251–259

[433] Redfern, M.: Stochastic differentiation - a generalized approach; *Acta Applicandae Mathematicae* **63** (2000) 349–361

[434] Redfern, M.: Complex white noise analysis; *Infinite Dimensional Analysis, Quantum Probability and Related Topics* **4** (2001) 1–29

[435] Redfern, M. and Betounes, D.: A generalized Itô formula for N-dimensional time; in: *White Noise Analysis*, World Scientific (1990) 337–343

[436] Saitô, K.: Itô's formula and Lévy's Laplacian; *Nagoya Math. J.* **108** (1987) 67–76

[437] Saitô, K.: Lévy's Laplacian in the infinitesimal generator; *Research Report, Meijo Univ.* **28** (1988) 1–5

[438] Saitô, K.: A computation of the Feynman integral in terms of complex white noise; *Research Reports of the Faculty of Science and Technology, Meijo University* **30** (1990) 13–18

[439] Saitô, K.: Itô's formula and Lévy's Laplacian II; *Nagoya Math. J.* **123** (1991) 153–169

[440] Saitô, K.: On a construction of a space of generalized functions; *Proc. PIC on Gaussian Random Fields, Part 2* (1991) 20–26

[441] Saitô, K.: A group generated by the Lévy Laplacian and the Fourier-Mehler transform; *Pitman Research Notes in Math. Series* **310** (1994) 274–288, Longman Scientific & Technical

[442] Saitô, K.: A group generated by the Lévy Laplacian; *RIMS Kokyuroku* **874** (1994) 192–201

[443] Saitô, K.: The Lévy Laplacian acting on Hida distributions; *Proc. Applied Math.* **5** (1996) 30–44

[444] Saitô, K.: The Fourier-Mehler transform and the Lévy Laplacian; *Infinite Dimensional Harmonic Analysis* **1** D.+M. Gräbner (1996) 195–208

[445] Saitô, K.: Transformations approximating a group generated by the Lévy Laplacian; *RIMS Kokyuroku* (1996) 214-224

[446] Saitô, K.: A (C_0)-group generated by the Lévy Laplacian; *Journal of Stochastic Analysis and Applications* **16** (1998) 567–584

[447] Saitô, K.: A (C_0)-group generated by the Lévy Laplacian II; *Infinite Dimensional Analysis, Quantum Probability and Related Topics* **1** (1998) 425–437

[448] Saitô, K.: The Lévy Laplacian and Stochastic Processes; *Infinite Dimensional Harmonic Analysis* **2**, D.+M. Gräbner (1999) 306–318

[449] Saitô, K.: Infinite dimensional stochastic processes generated by extensions of the Lévy Laplacian; *Publication of Centro Vito Volterra, Università Degli Studi di Roma "Tor Vergata"* **428** (2000) 1–12

[450] Saitô, K.: The Lévy Laplacian and Stochastic Processes; *RIMS Kokyuroku* **1157** (2000) 101–114

[451] Saitô, K.: The Lévy Laplacian and stable processes; in: *Chaos, Solitons & Fractals/ Les Treilles Special Issue*, International Solvay Institute (2001)

[452] Saitô, K.: A stochastic process generated by the Lévy Laplacian; *Acta Appl. Math.* (2001)

[453] Saitô, K., and Tsoi, A. H.: The Lévy Laplacian as a self-adjoint operator; *Quantum Information*, World Scientific (1999) 159–171

[454] Saitô, K., and Tsoi, A. H.: The Lévy Laplacian acting on Poisson noise functionals; *Infinite Dimensional Analysis, Quantum Probability and Related Topics* **2** (1999) 503–510

[455] Saitô, K., and Tsoi, A. H.: Stochastic processes generated by functions of the Lévy Laplacian; in: *Quantum Information II*, World Scientific (2000) 183–194

[456] Shieh, N. R. and Yokoi, Y.: Positivity of Donsker's delta function; *White Noise Analysis-Mathematics and Applications*, T. Hida et al. (eds.) (1990) 374–382, World Scientific

[457] Si Si: A note on Levy's Brownian motion II; *Nagoya Math. J.* **114** (1988) 166–172

[458] Si Si: Variational calculus for Lévy's Brownian motion; *Gaussian Random Fields*, K. Itô and T. Hida (eds.) (1991) 364–373, World Scientific

[459] Si Si: Innovation Approach to Levy's Brownian motion; *Proc. Preseminar for International Conference on Gaussian Random fields*, Part 2 (1991) 27–33

[460] Si Si: Integrability condition for stochastic variational equation; *Volterra Center Publications*, #217 (1995)

[461] Si Si: Innovation of some random fields; *J. Korean Math. Soc.* **35** (1998) 793–802.

[462] Si Si: A variation formula for some random fields, an analogy of Ito's formula; *Infinite Dimensional Analysis, Quantum Probability and Related Topics* **2** (1999) 305–313

[463] Si Si: Historical view, and some development of variational calculus applicable to random fields; *Volterra Center Publications*, #379 (1999)

[464] Si Si: Topics on random fields; *Quantum Information I*, T. Hida et al. (eds.), World Scientific, (1999) 179–194

[465] Si Si: Gaussian processes and Gaussian random fields; *Quantum Information II*, T. Hida et al. (eds.), World Scientific (2000) 195–204

[466] Si Si: Random fields and multiple Markov properties; *Supplementary Papers for Unconventional Model of Computation UMC'2K, Solvay Institute*, I. Antoniou et al. (eds.) (2000) 64–70

[467] Si Si: Random irreversible phenomena, entropy in subordination; *Chaos, Solitons & Fractals* (Special issue) I. Prigogine et al. (eds.) (2001)

[468] Si Si: Entropy in subordination and filtering; *Recent Developments in Infinite Dimensional Analysis and Quantum Probability*, L. Accardi et al. (eds.) (2001)

[469] Si Si: White noise approach to random fields; *Proc. Summer School on White Noise Approach to Classical and Quantum Stochastic Calculi*, L. Accardi (ed.), World Scientific (2001)

[470] Si Si: Representations and transformations of Gaussian random fields; *Quantum Information III*, T. Hida et al. (eds.), World Scientific (2001)

[471] Silva, J. L., Kondratiev, Yu. G., and Streit, L.: Representation of diffeomorphisms on compound Poisson space; in: *Analysis on Infinite-dimensional Lie Groups and Algebras* (1998) 376–393, World Scientific

[472] Skorokhod, A. V.: On a generalization of a stochastic integral; *Theory Probab. Appl.* **20** (1975) 219–233

[473] Stan, A.: Paley-Wiener theorem for white noise analysis; *J. Funct. Anal.* **173** (2000) 308–327

[474] Streit, L.: Gaussian processes and simple model field theories; in: *Proc. XII. Winter School for Theoretical Physics*, Karpacz, 1975

[475] Streit, L.: The construction of quantum field theories; in: *Uncertainty Principle and Foundations of Quantum Mechanics - A Fifty Years Survey*, S. S. Chissick (ed.), London, 1976

[476] Streit, L.: White noise analysis and the Feynman integral; in: *Functional Integration, Theory and Application*, J. P. Antoine and E. Tirapegui (eds.) Plenum (1980) 43–52

[477] Streit, L.: Energy forms, Schroedinger theory, processes; *Phys. Reports* **77** (1981) 363

[478] Streit, L.: Stochastic processes - quantum physics; *Acta Physica Austriaca Suppl., XXVI* (1984) 3

[479] Streit, L.: Quantum theory and stochastic processes - Some contact points; *XV. Conf. Stoch. Proc. Appl.*, Springer Lecture Notes in Mathematics **1203** (1985) 197

[480] Streit, L.: Energy forms in terms of white noise; in: *Stochastic, Algebra and Analysis in Classical and Quantum Dynamics*, S. Albeverio et al. (eds.), Kluwer (1990) 255–233

[481] Streit, L.: White noise analysis and quantum field theory; *Lecture Notes in Physics* **355** (1990) 287–298

[482] Streit, L.: The characterization theorem for Hida distributions. generalizations and applications; *UMa-Mat 3/91* and in: *Proc. III. Int. Conf. Stoch. Proc., Physics and Geometry*

[483] Streit, L.: White noise analysis - theory and applications; *Qu. Prob. Rel. Topics, VII* (1992) 337–347

[484] Streit, L.: The Feynman integral - recent results; in: *Dynamics of Complex and Irregular Systems*, Ph. Blanchard et al. (eds.), World Scientific (1993) 166–173

[485] Streit, L.: A new look at functional integration; in: *Advances in Dynamical Systems and Quantum Physics*, (1993) 307–325, World Scientific

[486] Streit, L.: White noise analysis and functional integrals; in: *Mathematical Approach to Fluctuations*, T. Hida (ed.), World Scientific

[487] Streit, L.: An Introduction to white noise analysis; in: *Stochastic Analysis and Applications in Physics*, A. I. Cardoso et al. (eds.), Kluwer (1994) 415–439

[488] Streit, L.: Hida distributions and more; in: *Mathematical Approach to Fluctuations 2*, T. Hida (ed.), World Scientific

[489] Streit, L.: The Feynman integral - answers and questions; in: *Proc. 1st Jagan Intl. Workshop on Adv. in Theor. Physics*, Central Visayan Inst. (1996) 188–199

[490] Streit, L.: Representations of diffeomorphisms on compound Poisson spaces; in: *Analysis on Infinite Dimensional Lie Algebras and Groups*, H. Heyer and J. Marion (eds.), (1998) World Scientific

[491] Streit, L. and Hida T.: Generalized Brownian functionals and the Feynman integral; *Stochastic Processes and Their Applications* **16** (1983) 55–69

[492] Streit, L. and Hida, T.: White noise analysis and its applications to Feynman integral; *Proc. Conf. Measure Theory and its applications, Lecture Notes in Math.* **1033** (1983) 219–226

[493] Streit, L. and Westerkamp, W.: A generalization of the characterization theorem for generalized functions of white noise; in: *Dynamics of Complex and Irregular Systems* Ph. Blanchard et al. (eds.), World Scientific (1993) 174–187

[494] Takenaka, S.: On projective invariance of multi-parameter Brownian motion; *Nagoya Math. J.* **67** (1977) 89–120

[495] Takenaka, S.: Invitation to white noise calculus; *Lecture Notes in Control and Inform. Sci.* **49** (1983) 249–257

[496] Takenaka, S.: On pathwise projective invariance of Brownian motion. I; *Proc. Japan Acad. Ser. A Math. Sci.* **64** (1988) 41–44

[497] Takenaka, S.: On pathwise projective invariance of Brownian motion. II; *Proc. Japan Acad. Ser. A Math. Sci.* **64** (1988) 271–274

[498] Takenaka, S.: On pathwise projective invariance of Brownian motion. III; *Proc. Japan Acad. Ser. A Math. Sci.* **66** (1990) 35–38

[499] Timpel, M. and Benth, F. E.: Topological aspects of the characterization of Hida distributions – a remark; *Stochastic Rep* **51** (1994) 293–299.

[500] Watanabe, H.: The local time of self-intersections of Brownian motions as generalized Brownian functionals; *Lett. Math. Phys.* **23** (1991) 1–9

[501] Watanabe, H.: Donsker's δ-function and its applications in the theory of white noise analysis; *Stochastic Processes*, a Festschrift in Honor of G. Kallianpur, S. Cambanis et al. (eds.) (1993) 337–339, Springer-Verlag

[502] Yan, J. A.: Sur la transformée de Fourier de H. H. Kuo; *Lecture Notes in Math.* **1372** (1989) 393–394, Springer-Verlag

[503] Yan, J. A.: Inequalities for products of white noise functionals; *Stochastic Processes* 349–358, Springer 1993

[504] Yan, J. A.: Products and transforms of white-noise functionals (in general setting); *Appl. Math. Optim.* **31** (1995) 137–153

[505] Yokoi, Y.: Positive generalized Brownian functionals; *White Noise Analysis, Mathematics and Applications*, World Scientific (1989) 407–422

[506] Yokoi, Y.: Positive generalized white noise functionals; *Hiroshima Math. J.* **20** (1990) 137–157

[507] Yokoi, Y.: Properties of Gel'fand triplet in white noise analysis and a characterization of Hida distributions; *Proc. Preseminar for International Conference on Gaussian Random Fields*, Part 2 (1991) 34–48

[508] Yokoi, Y.: Simple setting for white noise calculus using Bargmann space and Gauss transform; *Hiroshima Mathematical Journal* **25** (1995) 97–121

[509] Yokoi, Y.: On continuity of test functionals in infinite-dimensional Bargmann space; *Memoirs, Faculty of General Education, Kumamoto University, Natural Sciences* **31** (1996) 1–8

[510] Zhang, Y.: The Lévy Laplacian and Brownian particles in Hilbert spaces; *J. Funct. Anal.* **133** (1995) 425–441

Quantum Information IV (pp. 39–48)
Eds. T. Hida and K. Saitô

INTEGRAL TRANSFORM AND SEGAL-BARGMANN REPRESENTATION ASSOCIATED TO Q-CHARLIER POLYNOMIALS

NOBUHIRO ASAI

International Institute for Advanced Studies
Kizu, Kyoto 619-0225, Japan.
asai@iias.or.jp, nobuhiro.asai@nifty.com

Let $\mu_p^{(q)}$ be the q-deformed Poisson measure in the sense of Saitoh-Yoshida[24] and ν_p be the measure given by Equation (3.6). In this short paper, we introduce the q-deformed analogue of the Segal-Bargmann transform associated with $\mu_p^{(q)}$. We prove that our Segal-Bargmann transform is a unitary map of $L^2(\mu_p^{(q)})$ onto the q-deformed Hardy space $\mathcal{H}^2(\nu_q)$. Moreover, we give the Segal-Bargmann representation of the multiplication operator by x in $L^2(\mu_p^{(q)})$, which is the sum of the q-creation, q-annihilation, q-number, and scalar operators.

1 Introduction

The classical Segal-Bargmann transform in Gaussian analysis yields a unitary map of L^2 space of the Gaussian measure on $\mathbb{R}$ onto the space of L^2 holomorphic functions of the Gaussian measure on $\mathbb{C}$, see papers[7,8,15,16,19,25,26].

Recently, Accardi-Bożejko[1] showed the existence of a unitary operator between a one-mode interacting Fock space and L^2 space of a probability measure on $\mathbb{R}$ by making use of the basic properties of classical orthogonal polynomials and associated recurrence formulas[14,27]. Inspired by this work, the author[3] has recently extended the Segal-Bargmann transform to non-Gaussian cases. The crucial point is to introduce a coherent state vector as a kernel function in such a way that a transformed function, which is a holomorphic function on a certain domain in general, becomes a power series expression. Along this line, Asai-Kubo-Kuo[6] have considered the case of the Poisson measure compared with the case of the Gaussian measure. However, the case of L^2 space of Wigner's semi-circle distributions in free probability theory[28] is beyond their scope.

On the other hand, van Leeuwen-Maassen[22] considered a transform associated with q deformation of the Gaussian measure[10,11,12] and showed that for a given real number $q \in [0, 1)$ it is a unitary map of L^2 space of q-defomed Gaussian measure onto the q-deformed Hardy space $\mathcal{H}^2(\nu_q)$ where ν_q is given in (3.6). Biane[9] examined the case of $q = 0$ (Free case). Roughly speaking, their methods do not give the relationship between Szegö-Jacobi parameters and kernel functions for their transforms. As observed in Section 3 and Appendix A, our approach clarifies the relationship between them.

In this paper, we shall consider the q-deformed version of the Segal-Bargmann transform $S_{\mu_P^{(q)}}$ associated with q-deformed Poisson measure, denoted by $\mu_P^{(q)}$, in the sense of Saitoh-Yoshida[24]. As a main result, we shall provide Proposition 4.1, which claims that $S_{\mu_P^{(q)}}$ is a unitary map of $L^2(\mu_P^{(q)})$ onto $\mathcal{H}^2(\nu_q)$. Moreover, in Theorem 4.3 we shall give the representation in $\mathcal{H}^2(\nu_q)$ of the multiplication operator by x in $L^2(\mu_P^{(q)})$, which is the sum of the q-creation, q-annihilation, q-number, and scalar operators. We remark that our representation is compatible with that on the q-Fock space by Saitoh-Yoshida[23] and can be viewed as the q-analogue of the Hudson-Parthasarathy[17] decomposition of the usual Poisson random variable on the standard Boson Fock space $(q = 1)$. Ito-Kubo[18] also studied a similar decomposition in details from the point of white noise calculus[20,21] (For more recent formulation, see papers[4,5,13]).

The present article serves a good example to papers by Accardi-Bożejko[1] and Asai[3]. The present paper is organized as follows. In Section 2, we recall the recurrence formula for q-Charlier polynomials. In Section 3, we introduce a q-deformed coherent state vector and a Segal-Bargmann transform associated to a q-deformed Poisson measure. In addition, we quickly define the Hardy space as the Segal-Bargmann representation space. In Section 4, our main results are given. In Appendix A, we give some remarks on known results[7,9,22,25,26] related to q-Hermite polynomials.

Notation.

Let us recall standard notation from q-analysis[2]. We put for $n \in \mathbb{N}_0$,

$$[n]_q := 1 + q + \cdots + q^{n-1} \text{ with } [0]_q = 0.$$

Then q-factorial is naturally defined as

$$[n]_q! := [1]_q \cdots [n]_q \text{ with } [0]_q! = 0.$$

The q-exponential is given by

$$\exp_q(x) := \sum_{n=0}^{\infty} \frac{x^n}{[n]_q!}.$$

whose radius of convergence is $1/(1 - q)$. In addition, another symbol used is the q-analogue of the Pochhammer symbol,

$$(a;q)_n = \prod_{j=0}^{n-1} (1 - aq^j) \text{ and } (a;q)_\infty = \prod_{j=0}^{\infty} (1 - aq^j)$$

with the convention $(a;q)_0 = 1$.

2 q-deformed Charlier Polynomials

From now on, we always assume that $q \in [0,1)$ is fixed. Recently, Saitoh-Yoshida[24] calculated the explicit form of the q-deformed Poisson measure with a parameter $\beta > 0$ for $q \in [0,1)$. We denote it by $\mu_p^{(q)}$. The orthogonal polynomials associated to $\mu_p^{(q)}$ are the q-Charlier polynomials $\{C_n^{(q)}(x)\}$ with the Szegö-Jacobi parameters $\alpha_n = [n]_q + \beta, \omega_n = \beta[n]_q$. See papers[23,24]. We also refer the book[14] for the standard Charlier polynomials case $(q = 1)$. Let $\lambda = \{\beta^n[n]_q!\}_{n=0}^\infty$. The following relations hold for each $n \geq 1$:

$$(x - [n]_q - \beta)C_n^{(q)}(x) = C_{n+1}^{(q)}(x) + \beta[n]_q C_{n-1}^{(q)}(x) \tag{2.1}$$

where $C_0^{(q)} = 1$, $C_1^{(q)} = x - \beta$. Then, for any L^2-convergent decompositions $f(x) = \sum a_n C_n^{(q)}(x)$ and $g(x) = \sum b_n C_n^{(q)}(x)$, the inner product $\langle \cdot, \cdot \rangle_{L^2(\mu_p^{(q)})}$ is given by the form

$$\langle f, g \rangle_{L^2(\mu_p^{(q)})} = \sum_n \beta^n[n]_q! \bar{a}_n b_n. \tag{2.2}$$

Moreover, let $\|f\|_{L^2(\mu_p^{(q)})}^2 = \sum_{n=0}^\infty \beta^n[n]_q! |a_n|^2$.

3 Segal-Bargmann Transform and Hardy Space

Let us define the *q-deformed coherent state vector* for $\{C_n^{(q)}(x)\}$ by

$$E_{\lambda,p}^{(q)}(x,z) = \sum_{n=0}^\infty \frac{C_n^{(q)}(x)}{\beta^n[n]_q!} z^n, \quad z \in \Omega_q^\beta := \left\{ z \in \mathbb{C} : |z|^2 < \frac{\beta}{1-q} \right\}. \tag{3.1}$$

It is easy to see that $E_p^{(q)}(x,z) \in L^2(\mu_p^{(q)})$ due to

$$\|E_{\lambda,p}^{(q)}(x,z)\|_{L^2(\mu_p^{(q)})}^2 = \exp_q(\beta^{-1}|z|^2) < \infty. \tag{3.2}$$

Moreover, it can be shown that $\{E_{\lambda,p}^{(q)}(x,y)\}$ is linearly independent and total in $L^2(\mu_p)$.

Now we are in a position to introduce our key tool in this paper. Let us consider the q-analogue of the Segal-Bargmann transform $S_{\mu_P^{(q)}}$ associated to $\{C_n^{(q)}(x)\}$ given by

$$(S_{\mu_P^{(q)}} f)(z) = \langle E_{\lambda,p}^{(q)}(x,\bar z), f(x)\rangle_{L^2(\mu_P^{(q)})} \quad \text{for any } f \in L^2(\mu_P^{(q)}). \tag{3.3}$$

Lemma 3.1. *Let* $f \in L^2(\mu_P^{(q)})$. *Then* $(S_{\mu_P^{(q)}} f)(z)$ *converges absolutely for all* $z \in \Omega_q^\beta$.

Proof. For $f(x) = \sum_{n=0}^\infty a_n C_n^{(q)}(x)$, it is quite easy to see

$$(S_{\mu_P^{(q)}} f)(z) = \sum_{n=0}^\infty a_n z^n. \tag{3.4}$$

By the Schwartz inequality, we get the inequality

$$\sum_{n=0}^\infty |a_n z^n| \le \|f\|_{L^2(\mu_P^{(q)})} \exp_q\left(\frac{|z|^2}{\beta}\right).$$

This shows that the $(S_{\mu_P^{(q)}} f)(z)$ converges absolutely for all $z \in \Omega_q^\beta$. $\qquad\square$

The completion of the space of holomorphic functions F, G on Ω_q^β with respect to the inner product,

$$(F, G)_{\mathcal{H}_q^2} = \int \overline{F(z)} G(z) \nu_q(dz), \tag{3.5}$$

is nothing but the q-deformed Hardy space $\mathcal{H}^2(\nu_q)$. Here ν_q means that

$$\nu_q(dz) = (q;q)_\infty \sum_{j=0}^\infty \frac{q^j}{(q;q)_j} \lambda_{r_j}^\beta(dz), \quad q \in [0,1), \quad r_j = q^{\frac{j}{2}}\sqrt{\frac{\beta}{1-q}} \tag{3.6}$$

where $\lambda_{r_j}^\beta(dz)$ is the Lebesgue measure on the circle of radious r_j.

Lemma 3.2. $\{z^n\}_{n=0}^\infty$ *forms an orthogonal basis of* $\mathcal{H}^2(\nu_q)$.

Proof. We adopt the same idea as in the proof[22].

$$(z^n, z^m)_{\mathcal{H}_q^2} = \int_{\Omega_q^\beta} \overline{z}^n z^m \nu_q(dz)$$

$$= (q; q)_\infty \sum_{j=0}^\infty \frac{q^j}{(q; q)_j} \int_{\Omega_q^\beta} \overline{z}^n z^m \lambda_{r_j}^\beta (dz)$$

$$= \frac{(q; q)_\infty}{2\pi} \sum_{j=0}^\infty \frac{q^j}{(q; q)_j} r_j^{n+m} \int_0^{2\pi} e^{i(m-n)\theta} d\theta$$

$$= \delta_{n,m} \frac{\beta^n (q; q)_\infty}{(1 - q)^n} \sum_{j=0}^\infty \frac{q^{(n+1)j}}{(q; q)_j}$$

$$= \delta_{n,m} \beta^n [n]_q!.$$

Note that we have used the q-Gamma function[2,22],

$$\Gamma_q(n + 1) := \frac{(q; q)_\infty}{(1 - q)^n} \sum_{j=0}^\infty \frac{q^{(n+1)j}}{(q; q)_j} = [n]_q! \text{ for any } n \in \mathbb{N}.$$

$\square$

Remark. It can be shown by Proposition 4.4 in the recent paper[6] that ν_q is a unique measure satisfying $(z^n, z^m)_{\mathcal{H}_q^2} = \delta_{n,m} \beta^n [n]_q!$.

Hence, for any $F = \sum_{n=0}^\infty a_n z^n$, $G = \sum_{n=0}^\infty b_n z^n \in \mathcal{H}^2(\nu_q)$, the inner product $(\cdot, \cdot)_{\mathcal{H}_q^2}$ is written as

$$(F, G)_{\mathcal{H}_q^2} = \sum_{n=0}^\infty \beta^n [n]_q! \overline{a}_n b_n \tag{3.7}$$

and the corresponding norm of F is

$$\|F\|_{\mathcal{H}_q^2}^2 = \sum_{n=0}^\infty \beta^n [n]_q! |a_n|^2. \tag{3.8}$$

4   Main Results

Proposition 4.1. $S_{\mu_p^{(q)}}$ *is a unitary map of* $L^2(\mu_p^{(q)})$ *onto* $\mathcal{H}^2(\nu_q)$.

Proof. As we have seen in Lemma 3.1,

$$(S_{\mu_p^{(q)}} C_n^{(q)})(z) = z^n. \tag{4.1}$$

In addition, we derive by Lemma 3.2,

$$\|C_n^{(q)}\|^2_{L^2(\mu_p^{(q)})} = \|z^n\|^2_{\mathcal{H}_q^2} = \beta^n.$$

Therefore, we finish the proof. □

Let us define operators Z_q and D_q in $\mathcal{H}^2(\nu_q)$ satisfying

$$Z_q F(z) = z F(z). \tag{4.2}$$

and

$$D_{q,\beta} F(z) := 0 \ (z = 0), \quad D_{q,\beta} F(z) := \frac{\beta(F(z) - F(qz))}{z(1 - q)}. \ (z \neq 0) \tag{4.3}$$

Operators Z_q and $D_{q,\beta}$ play the roles of the *q-creation operator* and *q-annihilation operator* respectively and satisfy the q-deformed commutation relation $D_{q,\beta} Z_q - q Z_q D_{q,\beta} = I$. The *q-number operator* acting on $\mathcal{H}^2(\nu_q)$ is defined by

$$\tilde{N}_q F(z) = [n]_q F(z), \ n \geq 0. \tag{4.4}$$

In addition, the operator $\tilde{\alpha}_{N_q}$ acting on $\mathcal{H}^2(\nu_q)$ is defined by

$$\tilde{\alpha}_{N_q} F(z) = ([n]_q + \beta) F(z), \ n \geq 0. \tag{4.5}$$

Remark that $\tilde{\alpha}_{N_q} F(z) = 0$ for q-Gaussian case, see Appendix A. By the direct calculation, we have

Lemma 4.2. (1) $S_{\mu_p^{(q)}} 1 = 1$

(2) $D_{q,\beta} z^n = \beta[n]_q z^{n-1}$

(3) $Z_q z^n = z^{n+1}$

The transformation of the multiplication operator $Q_p^{(q)}$ by x in $L^2(\mu_p^{(q)})$ satisfies the following relation.

Theorem 4.3. $S_{\mu_p^{(q)}} Q_p^{(q)} = (D_{q,\beta} + Z_q + \tilde{\alpha}_{N_q}) S_{\mu_p^{(q)}}$

Proof. By the recurrence formula (2.1), Equation (4.1) and Lemma 4.2, we derive

$$(S_{\mu_p^{(q)}} Q_p^{(q)} C_n^{(q)})(z) = \left\langle E_{\lambda,p}^{(q)}(x,\bar{z}), C_{n+1}^{(q)} + \beta[n]_q C_{n-1}^{(q)} + ([n]_q + \beta) C_n^{(q)} \right\rangle_{L^2(\mu_p^{(q)})}$$

$$= (D_{q,\beta} + Z_q + \tilde{\alpha}_{N_q})(S_{\mu_p^{(q)}} C_n^{(q)})(z).$$

□

Moreover we obtain

Corollary 4.4. *The operators* $Z_q, D_q, \widetilde{N}_q, \widetilde{\alpha}_{N_q}$ *have the following properties:*
(1) $\widetilde{\alpha}_{N_q} = \frac{1}{\beta} Z_q D_{q,\beta} + \beta I = \widetilde{N}_q + \beta I$
(2) $Z_q + D_{q,\beta} + \widetilde{\alpha}_{N_q} = \left(\frac{1}{\sqrt{\beta}} Z_q + \sqrt{\beta} I \right) \left(\frac{1}{\sqrt{\beta}} D_{q,\beta} + \sqrt{\beta} I \right).$

Proof. The proof is done by Equations (4.2), (4.3), (4.4), (4.5), and Lemma 4.2. $\qquad\square$

Therefore, by Theorem 4.3 and Corollary 4.4, the multiplication operator $Q_p^{(q)}$ by x in $L^2(\mu_p^{(q)})$ is represented as the sum of the q-creation, q-annihilation, q-number, and scalar operators in the Hardy space $\mathcal{H}^2(\nu_q)$. In the q-deformed Gaussian case, due to Equation (A.4) in Appendix A, the multiplication operator $Q_g^{(q)}$ in $L^2(\mu_g^{(q)})$ is represented as a linear combination of q-creation and q-annihilation operators in the "same" Hardy space $\mathcal{H}^2(\nu_q)$.

Remark. (1) The analogous results in this paper to $q = 1$ have been considered by Asai, et al.[6]. (2) In general, if variances of two given measures μ_1 and μ_2 are the same, then the actions of creation and annihilation operators are the same. In addition, the transformed function spaces by S_{μ_1} and S_{μ_2} are also the same. However, if μ_1 is non-symmetric and μ_2 is symmetric, then the representation of multiplication in $L^2(\mu_1)$ is different from that in $L^2(\mu_2)$.

Appendix

A   On q-Hermite Polynomials

First of all, we refer to the papers[12,22] and references cited therein for the detailed description of q-Hermite polynomials. For $q = 1$, see books[14,21,27].

Let $\mu_g^{(q)}$ be the q-deformed Gaussian measure with mean zero and variance $\beta > 0$. It is well-known that an associated orthogonal polynomial to $\mu_g^{(q)}$ is the q-Hermite polynomial $\{H_n^{(q)}(x)\}$ with the Szegö-Jacobi parameters $\alpha_n = 0, \omega_n = \beta[n]_q$. In this case the following relations hold for each $n \geq 1$:

$$x H_n^{(q)}(x) = H_{n+1}^{(q)}(x) + \beta[n]_q H_{n-1}^{(q)}(x) \tag{A.1}$$

where $H_0^{(q)} = 1$ and $H_1^{(q)} = x$. Then for any $f(x) = \sum a_n H_n^{(q)}(x)$ and $g(x) = \sum b_n H_n^{(q)}(x)$ in $L^2(\mu_g^{(q)})$, the inner product $\langle \cdot, \cdot \rangle_{L^2(\mu_g^{(q)})}$ is given by the form

$$\langle f, g \rangle_{L^2(\mu_g^{(q)})} = \sum_n \beta^n [n]_q! \bar{a}_n b_n.$$

Moreover, let $\|f\|^2_{L^2(\mu_g^{(q)})} = \sum_{n=0}^{\infty} \beta^n [n]_q! |a_n|^2$.

A q-deformed coherent state vector $E_{\lambda,g}^{(q)}(x,z)$ for $\{H_n^{(q)}(x)\}$ is defined by

$$E_{\lambda,g}^{(q)}(x,z) = \sum_{n=0}^{\infty} \frac{H_n^{(q)}(x)}{\beta^n n!} z^n, \quad z \in \Omega_q^{\beta}. \tag{A.2}$$

It can be shown that the set $\{E_{\lambda,g}^{(q)}(x,z) : z \in \Omega_q^{\beta}\}$ is linearly independent and total in $L^2(\mu_g^{(q)})$. The q-analogue of the Segal-Bargmann transform $S_{\mu_g^{(q)}}$ associated to $\{H_n^{(q)}(x)\}$ is given by

$$(S_{\mu_g^{(q)}} f)(z) = \langle E_{\lambda,g}^{(q)}(x,\overline{z}), f(x) \rangle_{L^2(\mu_g^{(q)})} \text{ for any } f \in L^2(\mu_g^{(q)}). \tag{A.3}$$

With this transform, we can reproduce the same results as for $q \in [0,1)$ Theorem III.4 in Leeuwen-Maassen[22] and for $q = 0$ Proposition 1 in Biane[9]. That is, $S_{\mu_g^{(q)}}$-transform yields a unitary isomorphism between $L^2(\mu_g^{(q)})$ and $\mathcal{H}^2(\nu_q)$ and

$$S_{\mu_g^{(q)}} Q_g^{(q)} = (D_{q,\beta} + Z_q) S_{\mu_g^{(q)}} \tag{A.4}$$

where $Q_g^{(q)}$ is the multiplication operator by x in $L^2(\mu_g^{(q)})$.

Acknowledgments

The author is grateful for a Postdoctral Fellowship of the International Institute for Advanced Studies, Kyoto, Japan. He also thanks Prof. Bożejko for his comments.

References

1. L. Accardi and M. Bożejko, *Interacting Fock space and Gaussianization of probability measures.* Infinite Dimensional Analysis, Quantum Probability and Related Topics, **1** (1998), 663–670.
2. G. E. Andrew, R. Askey and R. Roy, *Special Functions.* Cambridge, 1999.
3. N. Asai: *Analytic characterization of one-mode interacting Fock space.* Infinite Dimensional Analysis, Quantum Probability and Related Topics, **4** (2001), 409–415.
4. N. Asai, I. Kubo, and H.-H. Kuo, *Roles of log-concavity, log-convexity, and growth order in white noise analysis.* Infinite Dimensional Analysis,

Quantum Probability and Related Topics, **4** (2001), 59–84. *Universidade da Madeira CCM preprint 37* (1999)

5. N. Asai, I. Kubo, and H.-H. Kuo, *General characterization theorems and intrinsic topologies in white noise analysis.* Hiroshima Math. J., **31** (2001), 299–330. *LSU preprint* (1999)

6. N. Asai, I. Kubo, and H.-H. Kuo, *Segal-Bargmann transforms of one-mode interacting Fock spaces associated to Gaussian and Poisson measures. preprint* (2001), to appear in *Proc. Amer. Math. Soc.*

7. V. Bargmann, *On a Hilbert space of analytic functions and an associated integral transform, I.* Comm. Pure Appl. Math., **14** (1961), 187–214.

8. V. Bargmann, *On a Hilbert space of analytic functions and an associated integral transform, II.* Comm. Pure Appl. Math., **20** (1967), 1–101.

9. P. Biane, *Segal-Bargmann transform, functional calculus on matrix spaces and the theory of semi-circular and circular systems.* J. Funct. Anal., **144** (1997), 232–286.

10. M. Bożejko and R. Speicher, *An example of a generalized Brownian motion I.* Comm. Math. Phys., **137** (1991), 519–531.

11. M. Bożejko and R. Speicher, *An example of a generalized Brownian motion II.* in Quantum Probability and Related Topics VII, L. Accardi (ed.) World Scientific, 1992, pp. 67–77.

12. M. Bożejko, B. Kümmerer and R. Speicher, *q-Gaussian processes: Noncommutative and classical aspects.* Comm. Math. Phys., **185** (1997), 129–154.

13. W. G. Cochran, H.-H. Kuo, and A. Sengupta, *A new class of white noise generalized functions.* Infinite Dimensional Analysis, Quantum Probability and Related Topics, **1** (1998), 43–67.

14. T. S. Chihara, *An Introduction to Orthogonal Polynomials.* Gordon and Breach, 1978.

15. T. Dwyer, *Partial differential equations in Fischer-Fock spaces for Hilbert-Schmidt holomorphy type.* Bull. Amer. Math. Soc. **77** (1971), 725–730

16. L. Gross P. and Malliavin, *Hall's transform and the Segal-Bargmann map.* in: Itô's Stochastic Calculus and Probability Theory, M. Fukushima et al. (eds.) Springer-Verlag, 1996, pp. 73–116.

17. R. L. Hudson and K. R. Parthasarathy, *Quantum Ito's formula and stochastic evolutions.* Comm. Math. Phys., **93** (1984), 301–323.

18. Y. Ito and I. Kubo, *Calculus on Gaussian and Poisson white noises.* Nagoya Math. J., **111** (1988), 41–84.

19. I. Kubo and H.-H. Kuo, *Finite dimensional Hida distributions.* J. Funct. Anal., **128** (1995), 1–47.

20. I. Kubo and S. Takenaka, Calculus on Gaussian white noise I, II, III, IV; *Proc. Japan Acad.* **56A** (1980) 376–380, **56A** (1980) 411–416, **57A** (1981) 433–437, **58A** (1982) 186–189.

21. H.-H. Kuo, *White Noise Distribution Theory.* CRC Press, 1996.

22. H. van Leeuwen and H. Maassen, *A q deformation of the Gauss distribution.* J. Math. Phys., **36** (1995), 4743–4756.

23. N. Saitoh and H. Yoshida, *The q-deformed Poisson random variables on the q-Fock space.* J. Math. Phys., **41** (2000), 5767–5772.

24. N. Saitoh and H. Yoshida, *The q-deformed Poisson distribution based on orthogonal polynomials.* J. Phys. A: Gen., **33** (2000), 1435–1444.

25. I. E. Segal, *Mathematical characterization of the physical vacuum for a linear Bose-Einstein field.* Illinois J. Math., **6** (1962), 500–523.

26. I. E. Segal, *The complex wave representation of the free Boson field.* in: Essays Dedicated to M. G. Krein on the Occassion of His 70th Birthday, Advances in Math.: Supplementary Studies Vol.3, I. Goldberg and M. Kac (eds.) Academic, 1978, pp. 321–344.

27. M. Szegö, *Orthogonal Polynomials.* Coll. Publ. **23**, Amer. Math. Soc., 1975.

28. D. Voiculescu, K. J. Dykema and A. Nica, *Free Rondom Variables.* CRM Monograph Ser. **1**, Amer. Math. Soc., 1992.

Quantum Information IV (pp. 49–78)
Eds. T. Hida and K. Saitô
© 2002 World Scientific Publishing Co.

NOTIONS OF INDEPENDENCE
IN ALGEBRAIC PROBABILITY THEORY
AND SET PARTITION STATISTICS

YUKIHIRO HASHIMOTO *

Graduate School of Mathematics, Nagoya University
Chikusa-ku, Nagoya 464-8602, Japan

A non-commutative analogy of the combinatorial probability theory is studied. We investigate roles of singletons in the set partition statistics which appear in the non-commutative probability theory, and introduce general notions of statistical independence. To clarify structures of the set partition statistics, we introduce an idea of non-commutative decomposition of discrete random walks, which is a sort of discrete time analogy of Quantum Ito theory.

1 Singleton condition and set partition statistics

An algebraic probability space $(\mathcal{A}, \varphi)$ is a pair of a $*$-algebra $\mathcal{A}$ and a state $\varphi : \mathcal{A} \to \mathbf{C}$. Let us take another $*$-algebra $\mathcal{B}$. A $*$-homomorphism $J : \mathcal{B} \to \mathcal{A}$ is called an *algebraic random variable* associated with a *sample algebra* $\mathcal{A}$ and a *state algebra* $\mathcal{B}$. Here and then we restrict ourselves in the category of unital $*$-algebras and then the algebraic random variables are unital $*$-homomorphism. A stochastic process on an algebraic probability space is a one-parameter family $\{J_t\}$ of algebraic random variables. Throughout this paper, we consider a stochastic process $\{J_n\}$ with discrete parameters $n \in \mathbf{N}$. As long as we fix a set of generators $\{b^{(j)}\}_{j \in \mathbf{N}} \subset \mathcal{B}$ it is sufficient to consider the set $\{a_n^{(j)} := J_n(b^{(j)}) \mid n, j \in \mathbf{N}\}$ instead of $\{J_n\}$. Thus we may identify $\{(a_n^{(j)})_{n=1}^{\infty} \mid j = 1, 2, \ldots\}$ with the discrete process.

Definition 1 [2,3]

(1) A finite or countably infinite set of sequences $\{(a_n^{(j)})_{n=1}^{\infty} \mid j = 1, 2, \ldots\}$ of elements in $\mathcal{A}$ is said to satisfy the *singleton condition* with respect to φ if the factorization

$$\varphi\left(a_{n_1}^{(j_1)} \cdots a_{n_m}^{(j_m)}\right) = \varphi(a_{n_s}^{(j_s)}) \cdot \varphi\left(a_{n_1}^{(j_1)} \cdots \hat{a}_{n_s}^{(j_s)} \cdots a_{n_m}^{(j_m)}\right) \tag{1}$$

occurs for any choice of $j_1, \ldots, j_m \in \mathbf{N}$ ($m \geq 1$), and $n_1, \ldots, n_m \in \mathbf{N}$ with an index n_s different from all other ones. Here $\hat{a}_{n_s}^{(j_s)}$ stands for the omission of $a_{n_s}^{(j_s)}$.

---

*JSPS RESEARCH FELLOW

(2) A stochastic process $\{J_n\}$ is said to satisfy the *singleton condition* with respect to φ if the factorization

$$\varphi\left(J_{n_1}(b^{(1)})\cdots J_{n_m}(b^{(m)})\right)$$
$$= \varphi(J_{n_s}(b^{(s)}))\cdot\varphi\left(J_{n_1}(b^{(1)})\cdots \hat{J}_{n_s}(b^{(s)})\cdots J_{n_m}(b^{(m)})\right) \quad (2)$$

holds for any choice of $b^{(1)},\ldots,b^{(m)} \in \mathcal{B}$, and $n_1,\cdots,n_m \in \mathbf{N}$ with an index n_s different from all other ones.

(3) We say sequences $(a_n^{(1)})_{n=1}^{\infty}, (a_n^{(2)}),\ldots$ of elements of $\mathcal{A}$ satisfy the condition of *boundedness of the mixed momenta* if for each $m \in \mathbf{N}$ and $j_1,\ldots,j_m$ there exists a positive constant $B_m(j_1,\ldots,j_m) > 0$ such that

$$\left|\varphi\left(a_{n_1}^{(j_1)}\cdots a_{n_m}^{(j_m)}\right)\right| \leq B_m(j_1,\ldots,j_m) \quad (3)$$

for any choice of $n_1,\ldots,n_m \in \mathbf{N}$.

Note that by setting $a_n^{(j)} := J_n(b^{(j)})$, the singleton condition for J_n implies (1), while the converse is false.

For a sequence $a^{(j)} = (a_n^{(j)})_{n=1}^{\infty} \subset \mathcal{A}$, we put

$$S_N(a^{(j)}) = \sum_{n=1}^{N} a_n^{(j)}. \quad (4)$$

We shall discuss the partition statistics associated with $\{S_N(a^{(j)})\}$. For describing the *mixed momenta* of $S_N(a^{(j)})$'s

$$\varphi\left(\frac{S_N(a^{(j_1)})}{N^{\alpha}}\cdots\frac{S_N(a^{(j_m)})}{N^{\alpha}}\right), \quad (5)$$

we introduce the following notations.

Notation 2 For $m \in \mathbf{N}$, we define a set of equivalent classes

$$P(m) = \{(W_1,\ldots,W_m) \mid 1 \leq W_i \leq m\}/\mathfrak{S}_m$$

where $(W_1, W_2,\ldots, W_m) \sim (W_1', W_2',\ldots, W_m')$ if and only if there exists a permutation $\theta \in \mathfrak{S}_m$ satisfying $\theta(W_i) = W_i'$. $[W_1\ldots W_m]$ stands for an equivalent class. For $T = [W_1\cdots W_m] \in P(m)$, $c(T)$ denotes the number of alphabets in T different from each other, that is, $c(T) := \#\{W_1,\ldots,W_m\}$. Then we may take a map $\rho : \{1,\ldots,m\} \to \{1,\ldots,c(T)\}$ characterized by $\rho(j) = \rho(k)$ if and only if $W_j = W_k$, which give rise to a partition $\coprod_{j=1}^{c(T)} S_j$ of the set $\{1,\ldots,m\}$ by putting $S_j := \{k \mid \rho(k) = j\}$. Since there is an ambiguity on the suffix of S_j, we choose the map ρ uniquely determined by the property $\min S_j < \min S_k$ if and only if $j < k$. $PP(2m)$ denotes the set of equivalent classes with length $2m$ which consist of m-pairs, that is, the set of all pair-partitions $\coprod_{j=1}^{m} S_j$ of the set $\{1,\ldots,2m\}$, $\#S_j = 2$ for $j = 1,\ldots,m$.

For an m-tuple $(n_1,\ldots,n_m) \in \mathbf{N}^m$, we construct a map $\rho' : \{1,\ldots,m\} \to \{1,\ldots,m\}$ defined by $\rho'(j) = \min\{k \mid n_k = n_j\}$, which has a property $\rho'(j) = \rho'(k)$ if and only if $n_j = n_k$. Then we associate an equivalent class $T = [\rho'(1)\cdots\rho'(m)]$ to each m-tuple $(n_1,\ldots,n_m)$, denoting $T = [(n_1,\ldots,n_m)]$ for short. We put for $T \in P(m)$,

$$T_N = \{(n_1,\ldots,n_m) \in \{1,\ldots,N\}^m \mid [(n_1,\ldots,n_m)] = T\}.$$

2 Law of large numbers, central limit theorems and entangled ergodicity

One can deduce some fundamental results from the singleton condition. Here we just mention the results and omit the proofs. See Accardi-Hashimoto-Obata [2,3] for details.

Lemma 3 [2,3] *Let $(a_n^{(1)})_{n=1}^{\infty}, (a_n^{(2)})_{n=1}^{\infty},\ldots$ be sequences of elements of $\mathcal{A}$ satisfying the condition of boundedness of the mixed momenta. Then, for any $0 < \alpha \leq 1$, $j_1,\ldots,j_m \in \mathbf{N}$ and $T \in P(m)$,*

$$\lim_{N\to\infty} \frac{1}{N^{\alpha m}} \sum_{(n_1,\ldots,n_m)\in T_N} \varphi\left(a_{n_1}^{(j_1)} \cdots a_{n_m}^{(j_m)}\right) = 0 \tag{6}$$

holds provided that $c(T) < \alpha m$.

Lemma 4 [2,3] *Let $\{(a_n^{(j)})_{n=1}^{\infty}\}$ be sequences of elements $\mathcal{A}$ with mean $\varphi(a_n^{(j)}) = 0$. Assume that the sequences satisfies the condition of mixed momenta (3) and the singleton condition with respect to φ. Then*

(i) if $\alpha > 1/2$ or if $\alpha = 1/2$ and m is odd, we have

$$\lim_{N\to\infty} \varphi\left(\frac{S_N(a^{(j_1)})}{N^{\alpha}} \cdots \frac{S_N(a^{(j_m)})}{N^{\alpha}}\right) = 0. \tag{7}$$

(ii) In the case of $\alpha = 1/2$ and $m = 2p$,

$$\lim_{N\to\infty} \varphi\left(\frac{S_N(a^{(j_1)})}{\sqrt{N}} \cdots \frac{S_N(a^{(j_{2p})})}{\sqrt{N}}\right)$$
$$= \lim_{N\to\infty} \frac{1}{N^p} \sum_{T\in PP(2p)} \sum_{(n_1,\ldots,n_{2p})\in T_N} \varphi\left(a_{n_1}^{(j_1)} \cdots a_{n_{2p}}^{(j_{2p})}\right) \tag{8}$$

holds in the sense that one limit exists if and only if the other does and the limits coincide. Moreover, the following Gaussian bound takes place:

$$\limsup_{N \to \infty} \left| \varphi \left(\frac{S_N(a^{(j_1)})}{\sqrt{N}} \cdots \frac{S_N(a^{(j_{2p})})}{\sqrt{N}} \right) \right| \leq \frac{(2p)!}{2^p p!} B_{2p}. \tag{9}$$

Definition 5 [5]   A discrete stochastic process $J_n : \mathcal{B} \to \mathcal{A}$ is said to be *weakly stationary* with respect to φ if the distribution $\psi = \varphi \circ J_n$ is determined independent of n.

The following is a non-commutative version of the law of large numbers.

Theorem 6 [5]   *Let $(\mathcal{A}, \varphi)$ be an algebraic probability space and $J_n : \mathcal{B} \to \mathcal{A}$ be a stochastic process. Suppose that $\{J_n\}$ is weakly stationary and satisfies the singleton condition (2) with respect to φ. Moreover, suppose that for any $m \in \mathbf{N}$ and any choice of $b^{(1)}, \ldots, b^{(m)} \in \mathcal{B}$, there exists a constant $B_m(b^{(1)}, \ldots, b^{(m)}) > 0$ such that*

$$|\varphi(J_{n_1}(b^{(1)}) \cdots J_{n_m}(b^{(m)}))| \leq B_m(b^{(1)}, \ldots, b^{(m)}), \qquad n_1, \ldots, n_m \in \mathbf{N} \tag{10}$$

holds. Then for any $m \in \mathbf{N}$, any m-variable non-commutative polynomial P and any choice of $b^{(1)}, \ldots, b^{(m)} \in \mathcal{B}$,

$$\lim_{N \to \infty} \varphi \left(P \left(\frac{S_N(a^{(1)})}{N}, \ldots, \frac{S_N(a^{(m)})}{N} \right) \right) = P(\psi(b^{(1)}), \ldots, \psi(b^{(m)})), \tag{11}$$

where $a_n^{(j)} = J_n(b^{(j)})$ and $\psi = \varphi \circ J_n$.

In the next, we observe the natural role of the non-crossing partitions in the proof of the existence of the limit (8) and show how this naturally leads to the idea of *entangled ergodicity*.

Definition 7 Let $\coprod_{j=1}^{k} S_j$ be a partition of $\{1, \ldots, m\}$ and put

$$\underline{s}_j = \min\{s \in S_j\}, \qquad \overline{s}_j = \max\{s \in S_j\}.$$

Then $\coprod_{j=1}^{k} S_j$ is called *non-crossing* if for any $i, j \in \{1, \ldots, m\}$ distinct each other,

$$(\underline{s}_i, \overline{s}_i) \cap S_j = \emptyset \quad \text{or} \quad (\underline{s}_j, \overline{s}_j) \cap S_i = \emptyset \tag{12}$$

holds. Here (a, b) stands for the set $\{a, a+1, \ldots, b\}$. The block S_j is called a *non-crossing block* of a partition if (12) holds for any S_i, $i \neq j$. The partition $\coprod_{j=1}^{k} S_j$ is called *totally crossing* if there is no non-crossing block.

Suppose that the process $\{J_n\}$ satisfies the singleton condition. Then, for non-crossing pair partition $T = \coprod_{k=1}^{p}\{h_k, l_k\} \in PP(2p)$, the factorization property (2) shows that the limit

$$\lim_{N\to\infty} \frac{1}{N^p} \sum_{(n_1,\ldots,n_{2p})\in T_N} \varphi(J_{n_1}(b^{(j_1)})\cdots J_{n_{2p}}(b^{(j_{2p})})) \tag{13}$$

is reduced to the limit of usual ergodic average:

$$\lim_{N\to\infty} \varphi\left(\frac{1}{N}\sum_{n=1}^{N} J_n(b^{(h_1)}b^{(l_1)})\right) \cdots \varphi\left(\frac{1}{N}\sum_{n=1}^{N} J_n(b^{(h_p)}b^{(l_p)})\right).$$

However, due to the non-commutativity we need the existence of the limit for entangled partitions.

Definition 8 [2,3] A process $\{J_n\}$ is said to have the property of *entangled ergodicity* with respect to φ if for any $m \in \mathbf{N}$, any choice of $b^{(1)},\ldots,b^{(m)} \in \mathcal{B}$ and any totally crossing partition $T \in P(m)$, $c(T) \geq 2$, the limit

$$\lim_{N\to\infty} \frac{1}{N^{c(T)}} \sum_{(n_1,\ldots,n_m)\in T_N} \varphi(J_{n_1}(b^{(1)})\cdots J_{n_m}(b^{(m)})) \tag{14}$$

exists.

Theorem 9 [2,3] *Suppose that a process $\{J_n\}$ satisfies the singleton condition. Let $(b^{(j)})$ be a sequence of $\mathcal{B}$ with $\varphi(J_n(b^{(j)})) = 0$ for any n and j, satisfying uniformly bounded condition (10). Put $a_n^{(j)} := J_n(b^{(j)})$. Suppose that the mean covariance*

$$\lim_{N\to\infty} \frac{1}{N} \sum_{n=1}^{N} \varphi(J_n(b^{(i)}b^{(j)})) \tag{15}$$

exists for any i, j. Then the limit

$$\lim_{N\to\infty} \varphi\left(\frac{S_N(a^{(1)})}{\sqrt{N}} \cdots \frac{S_N(a^{(2p)})}{\sqrt{N}}\right) \tag{16}$$

exists if and only if $\{J_n\}$ satisfies the entangled ergodicity.

Corollary 10 [2,3] *Suppose that a process $\{J_n\}$ satisfies the singleton condition. Let $(b^{(j)})$ be a sequence of $\mathcal{B}$ with $\varphi(J_n(b^{(j)})) = 0$ for any n and j, satisfying uniformly bounded condition (10). Put $a_n^{(j)} := J_n(b^{(j)})$. The central limit theorem holds if any one of the following conditions is satisfied:*

(i) (q–commutation relations) for each $i, j \in \mathbf{N}$, $i \neq j$, there exists a complex number q_{ij} such that $a_m^{(i)} a_n^{(j)} = q_{ij} a_n^{(j)} a_m^{(i)}$ for any $m, n \in \mathbf{N}$;

54

(ii) $(symmetry)$

$$\varphi(a_{n_1}^{(j_1)}\cdots a_{n_m}^{(j_m)}) = \varphi(a_{\theta(n_1)}^{(j_1)}\cdots a_{\theta(n_m)}^{(j_m)}), \tag{17}$$

holds for any injection $\theta : \mathbf{N} \to \mathbf{N}$,

(iii) *(pair partition freeness)* $\varphi(a_{n_1}^{(j_1)}\cdots a_{n_{2p}}^{(j_{2p})}) = 0$, $(n_1,\ldots,n_{2p}) \in T$ *for any totally crossing pair partition $T \in PP(2p)$ and any of $j_1,\ldots,j_{2p}$.*

The existence of the limit (14) on the totally crossing partition is proved by Fichitner-Freudenberg-Liebscher[6] for weak stationary processes. The validity of the entangled ergodic theorem would imply that the usual stationarity condition is sufficient to guarantee the validity of the central limit theorem under the only assumption of the singleton condition.

3 Set partition statistics on discrete groups

Let G be a discrete group generated by $\Sigma := \{\sigma_n|\ n \in \mathbf{N}\}$. Throughout this section we assume all σ_i's are of order 2, that is, $\sigma_n^{-1} = \sigma_n$ for simplicity. This assumption is not essential for a study of relations between partition statistics and group relations. See the remark (1) at the end of this section. Let $\pi : G \to \mathbf{B}(l^2(G))$ be the left regular representation. Let $\mathcal{A}_G$ be the $*$-algebra generated by $\pi(G)$ with $*$-operation $(a\pi(g) + b\pi(h))^* = \bar{a}\pi(g^{-1}) + \bar{b}\pi(h^{-1})$ $(a,b \in \mathbf{C}, g,h \in G)$. In this section we take a vacuum state $\phi(g) := \langle 0|\pi(g)|0\rangle$ on $\mathcal{A}_G$ where $|0\rangle \in l^2(G)$ stands for the characteristic function δ_e of the unit $e \in G$ and consider the pair $(\mathcal{A}_G, \phi)$ as an algebraic probability space.

The $*$-linear extension of a group homomorphism $J_n : \mathbf{Z}/2\mathbf{Z} \ni \sigma \mapsto \sigma_n \in G$ is taken for an algebraic random variable associated with a sample algebra $\mathcal{A}_G$ and state algebra $\mathcal{B} := \langle \mathbf{Z}/2\mathbf{Z}\rangle_{*\text{-algebra}}$. An algebraic central limit theorem describes a limit behavior of rescaled sum of algebraic random variables with mean 0 and variance 1,

$$S_N = \frac{1}{\sqrt{N}}(\pi(\sigma_1) + \ldots + \pi(\sigma_N)).$$

Note that a family of subsets $\Sigma^{(N)} := \{\sigma_n \mid n = 1,\ldots,N\} \subset \Sigma$ gives a filtration of G: $G^{(1)} \subset G^{(2)} \subset \cdots \subset G$, where $G^{(N)}$ is the subgroup generated by $\Sigma^{(N)}$. Then S_N is the graph Laplacian acting on a Cayley graph defined by $\Sigma^{(N)}$. We are interested in limit distribution of S_N under the vacuum state. Let us consider the following limit

$$M_m = M_m(G, \Sigma) = \lim_{N\to\infty} \phi(S_N^m)$$

$$= \lim_{N \to \infty} \frac{1}{(\sqrt{N})^m} \sum_{1 \le i_1, \ldots, i_m \le N} \phi(\sigma_{i_1} \sigma_{i_2} \cdots \sigma_{i_m})$$

$$= \lim_{n \to \infty} \frac{1}{(\sqrt{N})^m} \sum_{T_N \in P_G(m)} \#T_N.$$

We call $\{M_m\}$ *combinatorial moments*. Let Δ_k be a $k \times k$-matrix given by $(\Delta_k)_{ij} = M_{i+j}$, $0 \le i + j \le k - 1$. It is known that the Hamburger's problem is solved if and only if Δ_k is non-negative definite for all k. In addition to this condition, if M_m's satisfy Carleman's condition

$$\limsup_{m \to \infty} \frac{1}{2m} M_{2m}^{1/2m} < \infty, \tag{18}$$

then there exists a unique distribution μ_G on real line of which moments coincide with $\{M_m\}$.

For a word $(\sigma_{i_1}, \cdots, \sigma_{i_m}) \in \Sigma^m$, we define a map $\rho : \{1, \ldots, m\} \to \{1, \ldots, m\}$ by $\rho(j) = \min\{k \mid i_k = i_j\}$. Note that $\rho(j) = \rho(k)$ if and only if $i_j = i_k$. Then we associate an equivalent class $T = [\rho(1) \cdots \rho(m)]$ to each word $(\sigma_{i_1}, \ldots, \sigma_{i_m})$. We use an abbreviation $T = [\sigma_{i_1} \cdots \sigma_{i_m}]$. If a product $\sigma_{i_1} \cdots \sigma_{i_m}$ is reduced to the unit e, we say the word $(\sigma_{i_1}, \ldots, \sigma_{i_m})$ is *closed*. $P_G(m)$ denotes the subset of $P(m)$ which consists of equivalent classes of closed words in (G, Σ). We put $PP_G(2m) = P_G(2m) \cap PP(2m)$.

Definition 11 A set of generators Σ of G is called *minimal* if no proper subset of Σ generates G.

Proposition 12 [7,4] *If Σ is minimal, then Σ satisfies the singleton condition with respect to the vacuum state ϕ.*

PROOF. If a product $\sigma_{i_1} \cdots \sigma_{i_k}$ includes a singleton σ_{i_s}, we have $\sigma_{i_1} \cdots \sigma_{i_k} \ne e$. Otherwise we have $\sigma_{i_s} = \sigma_{i_{s-1}} \cdots \sigma_{i_1} \cdot \sigma_{i_k} \cdots \sigma_{i_{s+1}}$, which contradicts to the minimality of Σ. Then we obtain $\phi(\sigma_{i_1} \cdots \sigma_{i_k}) = 0$. It follows the singleton condition from $\phi(\sigma_{i_s}) = 0$. ∎

Definition 13 A set of generators $\{\sigma_i\}$ is called *symmetric* if a word $(\sigma_{i_1}, \ldots, \sigma_{i_k})$ is closed, so is $(\sigma_{\theta(i_1)}, \ldots, \sigma_{\theta(i_k)})$ for any injection $\theta : \mathbf{N} \to \mathbf{N}$. If the generators are symmetric then one has $P_{G^{(N)}}(m) = P_G(m)$ for $N \ge m$ and the combinatorial moments $\{M_m\}$ are given by

$$M_m = \lim_{n \to \infty} \sum_{T \in P_G(m)} \frac{1}{(\sqrt{N})^m} \#T_N \tag{19}$$

Theorem 14 [7] *Let G be a discrete group generated by symmetric minimal generators $\{\sigma_i\}$. Then all combinatorial moments M_m exist and are given by*

$$M_{2m} = \#PP_G(2m) \quad and \quad M_{2m+1} = 0. \tag{20}$$

56

PROOF. Take a closed word $(\sigma_{i_1}, \ldots, \sigma_{i_m})$, then $\phi(\sigma_{i_1} \cdots \sigma_{i_m}) = 1$ holds. Since Σ is minimal, it follows from Lemma 4 and the symmetry assumption that for odd m we have $M_m = 0$, and for even $m = 2p$ the right hand side of (19) equals to

$$\lim_{n \to \infty} \sum_{T \in PP_G(2p)} \frac{1}{N^p} \#T_N = \#PP_G(2p).$$

■

Here are two typical results in algebraic probability theory.

- Let A be a direct sum of $\mathbf{Z}/2\mathbf{Z}$ with minimal symmetric generators: $A = \oplus_{i=1}^{\infty} \mathbf{Z}/2\mathbf{Z}e_i$. It is easy to see that $PP_A(2m) = PP(2m)$, and hence

$$M_{2m} = \frac{(2m)!}{2^m m!} \quad \text{and} \quad M_{2m+1} = 0.$$

Then the central limit distribution for $(A, \{e_i\})$ under the vacuum state ϕ equals to the normalized Gaussian distribution

$$d\mu_A(x) = \frac{1}{\sqrt{2\pi}} e^{-\frac{1}{2}x^2} dx.$$

- Let F be a free product of $\mathbf{Z}/2\mathbf{Z}$, $F = *_{i=1}^{\infty} \mathbf{Z}/2\mathbf{Z}f_i$. It is shown that $PP_F(2m)$ is a set of non-crossing pair partitions of $\{1, \ldots, 2m\}$, and hence

$$M_{2m} = \#PP_F(2m) = \frac{(2m)!}{(m+1)!m!} \quad \text{and} \quad M_{2m+1} = 0.$$

(See e.g. [14].) Then the central limit distribution for $(F, \{f_i\})$ under the vacuum state ϕ equals to the normalized semi-circle distribution

$$d\mu_F(x) = \frac{1}{2\pi} \chi_{[-2,2]} \sqrt{4 - x^2} dx$$

We shall discuss these examples again in section 8.1 from the viewpoint of distance regular graphs. Here we only note that these examples are extremal in the discrete group category.

Theorem 15 [7,4] *Let G be a discrete group generated by symmetric minimal generators $\{\sigma_i\}$. Then the moments of limit distribution satisfy*

$$\frac{(2m)!}{(m+1)!m!} \leq M_{2m}(G, \{\sigma_i\}) \leq \frac{(2m)!}{2^m m!}. \tag{21}$$

Therefore the sequence $\{M_m(G, \{\sigma_i\})\}$ satisfies the Carleman's uniqueness condition.

PROOF. By the universal property of free groups, there exists a canonical surjection $\pi : (F, \{f_i\}) \twoheadrightarrow (G, \Sigma)$ with $\pi(f_i) = \sigma_i$. Then $PP_F(2m) \subset PP_G(2m)$ holds for all m, hence $(2m)!/(m+1)!m! = \#PP_F(2m) \leq \#PP_G(2m) = M_{2m}(G)$. On the other hand, $PP_G(2m)$ is a subset of $PP(2m)$ and one sees $\#PP(2m) = (2m)!/2^m m!$. Hence the inequality follows. Since the moments $\{(2m)!/2^m!m!\}$ fulfill the condition (18), the rest of the assertion follows. $\blacksquare$

Remarks.

(1) The results in this section are valid with slight modification in case where all generators are of order more than 2 (*cf.* Hashimoto[7]). In that case, one needs to consider $\{S_N\}$ as

$$S_N = \frac{1}{\sqrt{2N}}(\pi(\sigma_1) + \pi(\sigma_1^{-1}) + \cdots + \pi(\sigma_N) + \pi(\sigma_N^{-1})).$$

Then a 'pair' in a product means that $\sigma_{i_1}^{\varepsilon_1} \cdots \sigma_{i_m}^{\varepsilon_m}$ is given by $(\sigma_{i_p}^{\varepsilon_p}, \sigma_{i_q}^{\varepsilon_q})$ with $i_p = i_q$ and $\varepsilon_p = -\varepsilon_q$. The free abelian group $A = \oplus^\infty \mathbf{Z}$ and the free group $F = *^\infty \mathbf{Z}$ play the same extremal role as stated in Theorem 15.

(2) The concept of minimal generator is extended to the category of monoids, see Accardi-Hashimoto-Obata[4] for details.

(3) Lu[15,16] and Muraki[17] independently introduced a new notion of independence in algebraic probability theory, called *chronological* or *monotone* independence, where the symmetric condition on 'generators' breaks down. Then we fail to apply our result to a sequence of monotone independent random variables. Indeed, the limit distribution associated with a sum of monotone independent random variables with respect to a vacuum state coincides with the arcsine distribution, of which combinatorial moments are determined by a certain set of pair partitions which have much smaller cardinality than the set of non-crossing pair partitions.

(4) In general, we face hard combinatorial problems to determine combinatorial moments associated with $(G, \{\sigma_i\})$. If we construct a suitable space on which G acts, there is a possibility to determine combinatorial moments by observing the action of G. We illustrate such a method in the following example. We forget the order 2 assumption on generators here.

For a fixed $k \in \mathbf{N}$, let $\mathcal{N}^{(k)}$ be the normal subgroup of infinite free group $F = *_{i=1}^\infty \mathbf{Z} f_i$ generated by $\{(f_{i_1} f_{i_2} \cdots f_{i_l})^{kl+1} | l, i_j \in \mathbf{N}\}$. We put $\mathfrak{S}^{(k)} = F/\mathcal{N}^{(k)}$ and $\tau_i = f_i \mathcal{N}^{(k)}$. We note that the infinite permutation group $\mathfrak{S}_\infty$ is a factor group of $\mathfrak{S}^{(2)}$.

Proposition 16 [7] *The generators $\{\tau_i\}$ are symmetric and minimal, and*

$$PP_{\mathfrak{S}^{(k)}}(2m) = PP_F(2m).$$

Then, the central limit distribution associated with $(\mathfrak{S}^{(k)}, \{\tau_i\})$ coincide with the normalized semi-circle law.

PROOF.   Let X be a discrete space $\{0\} \cup \mathbf{Z} \times \{1, 2, \ldots, k\}$. Define the action of τ_i's on X by

$$\tau_i(0) = (i, 1), \qquad \tau_i(i, k) = 0,$$
$$\tau_i(i, \alpha) = (i, \alpha + 1), \quad \text{for } 1 \le \alpha \le k - 1,$$
$$\tau_i(j, \alpha) = (j, \alpha), \quad \text{for } j \ne i,$$

and τ_i^{-1}'s on X inverse way. We see that $(\tau_{i_1} \tau_{i_2} \cdots \tau_{i_l})^{kl+1}$ acts on X identically for every $l \in \mathbf{N}$. Then the action of $\mathfrak{S}^{(k)}$ on X is well defined. Suppose $\tau_{i_1}^{\varepsilon_1} \cdots \tau_{i_s}^{\varepsilon_s} = e$ and that τ_{i_p} is a singleton. Then $\tau_{i_1}^{\varepsilon_1} \cdots \tau_{i_{p-1}}^{\varepsilon_{p-1}}$ and $\tau_{i_{p+1}}^{\varepsilon_{p+1}} \cdots \tau_{i_s}^{\varepsilon_s}$ leave (i_p, α) fixed, while $\tau_{i_p}^{\varepsilon_p}$ changes (i_p, α). This is a contradiction. Hence the minimality of $\{\tau_i\}$ follows. By Theorem 14 we focus on a closed word $(\tau_{i_1}^{\varepsilon_1}, \cdots, \tau_{i_{2m}}^{\varepsilon_{2m}})$ belongs to a pair partition. Choose a pair (τ_{i_p}, τ_{i_q}) $(p < q)$, i.e., $r := i_p = i_q$ and $\varepsilon_q = -\varepsilon_p$, such that $\tau_{i_{p+1}}, \ldots, \tau_{i_{q-1}}$ are different each other. We assume $\varepsilon_q = 1$ without losing generality. Then we have $\tau_{i_p}^{\varepsilon_p} \cdots \tau_{i_q}(r, k) = (i_{q-1}, 1)$ and it follows that

$$\tau_{i_1}^{\varepsilon_1} \cdots \tau_{i_{2m}}^{\varepsilon_{2m}}(r, k) = \tau_{i_1}^{\varepsilon_1} \cdots \tau_{i_q}(r, k)$$
$$= \tau_{i_1}^{\varepsilon_1} \cdots \tau_{i_{p-1}}^{\varepsilon_{p-1}}(i_{q-1}, 1)$$
$$\ne (r, k)$$

whenever $q \ne p+1$. By the assumption $\tau_{i_1}^{\varepsilon_1} \cdots \tau_{i_{2m}}^{\varepsilon_{2m}} = e$, we need $q = p+1$. Repeat this argument, then we see that the closed word $(\tau_{i_1}^{\varepsilon_1}, \cdots, \tau_{i_{2m}}^{\varepsilon_{2m}})$ forms a non-crossing pair partition, and hence the assertion follows.   ∎

In the argument above, the action of $\mathfrak{S}^{(k)}$ on X is used to exclude all kinds of pair partitions except non-crossing ones. This method is useful to determine the moments.

4   Singleton independence

Motivated by the study of central limit theorem for the Haagerup states on free groups [8], we introduce an analytic generalization of the singleton condition as follows.

Definition 17 [2,3] Let $\mathcal{A}$ be a $*$-algebra and let $(a_n^{(j)})_{n=1}^{\infty}$ be sequences of $\mathcal{A}$. Assume that we are given a family of states φ_γ, $\gamma \geq 0$, on $\mathcal{A}$. The sequences $(a_n^{(j)})$ is called *singleton independent* with respect to φ_γ if for any $m \in \mathbf{N}$ and any choice of $j_1, \ldots, j_m$, there exists a constant $C_m := C_m(j_1, \ldots, j_m) > 0$ such that

$$|\varphi_\gamma(a_{n_1}^{(j_1)} \cdots a_{n_m}^{(j_m)})| \leq C_m \left|\varphi_\gamma(a_{n_s}^{(j_s)})\right| \cdot \left|\varphi_\gamma\left(a_{n_1}^{(j_1)} \cdots \hat{a}_{n_s}^{(j_s)} \cdots a_{n_m}^{(j_m)}\right)\right| \quad (22)$$

holds whenever n_s is a singleton for $(n_1, \ldots, n_m)$.

An algebraic random process $J_n : \mathcal{B} \to \mathcal{A}$ is called *singleton independent* with respect to φ_γ if for any $m \in \mathbf{N}$ and any choice of $b^{(j_1)}, \ldots, b^{(j_m)} \in \mathcal{B}$, there exists a constant $C_m := C_m(b^{(j_1)}, \ldots, b^{(j_m)}) > 0$ such that (22) holds for $a_n^{(j)} := J_n(b^{(j)})$ whenever n_s is a singleton for $(n_1, \ldots, n_m)$.

We assume that there exist constant numbers $\beta > 0$, $\kappa > 0$ and $\lambda \geq 0$ such that

$$\sup_{j,n \in \mathbf{N}} |\varphi_\gamma(a_n^{(j)})| \leq \gamma^\beta. \quad (23)$$

Then in the case of $\gamma = 0$, the condition (22) is reduced to the singleton condition. (Here we put $0^0 = 1$.) Our goal is to study the partition statistics in the limit associated with mixed momenta

$$\lim_{N \to \infty} \varphi_{\gamma(N)} \left(\frac{S_N(a^{(1)})}{N^\alpha} \cdots \frac{S_N(a^{(m)})}{N^\alpha}\right)$$

$$= \lim_{N \to \infty} \frac{1}{N^{\alpha m}} \sum_{T \in P(m)} \sum_{(n_1, \ldots, n_m) \in T_N} \varphi_{\gamma(N)} \left(a_{n_1}^{(1)} \cdots a_{n_m}^{(m)}\right), \quad (24)$$

where $\gamma(N)$ is a function of N satisfying

$$\lim_{N \to \infty} \gamma(N) N^\kappa = \lambda. \quad (25)$$

Lemma 18 *Let $s(T)$ denote the number of singletons in the partition $T \in P(m)$. It holds that for any partition $T \in P(m)$,*

$$m + s(T) \geq 2c(T), \quad (26)$$

and the equality holds if and only if T consists of pairs and singletons.

PROOF. For a partition $T = \coprod_{j=1}^{c(T)} S_j \in P(m)$, we have

$$m = \sum_{j=1}^{c(T)} \#S_j = s(T) + \sum_{\#S_j \geq 2} \#S_j \geq 2(c(T) - s(T)) + s(T) = 2c(T) - s(T).$$

The equality holds if and only if $\#S_j = 2$ for all j. $\blacksquare$

Lemma 19 *Suppose that sequences $(a_n^{(j)})$ satisfy the condition of boundedness of mixed momenta (3) uniformly on γ, and are singleton independent. If a partition $T \in P(m)$ satisfies*

$$\alpha m + \beta \kappa s(T) > c(T) \tag{27}$$

then we have the limit

$$\lim_{N \to \infty} \frac{1}{N^{\alpha m}} \sum_{(n_1, \ldots, n_m) \in T_N} \varphi_{\gamma(N)} \left(a_{n_1}^{(1)} \cdots a_{n_m}^{(m)} \right) = 0.$$

PROOF. The boundedness (3) and repeated application of (22) yields an inequality

$$\left| \varphi_{\gamma(N)} \left(a_{n_1}^{(1)} \cdots a_{n_m}^{(m)} \right) \right| \le D_m \lambda^{\beta s(T)} N^{-\beta \kappa s(T)},$$

where $D_m := B_{m-s(T)} C_m C_{m-1} \cdots C_{m-s(T)+1}$. Hence we have

$$\frac{1}{N^{\alpha m}} \left| \sum_{(n_1, \ldots, n_m) \in T_N} \varphi_{\gamma(N)} \left(a_{n_1}^{(1)} \cdots a_{n_m}^{(m)} \right) \right| \le D_m \frac{\lambda^{\beta s(T)} (c(T))!}{N^{\alpha m + \beta \kappa s(T)}} \binom{N}{c(T)}.$$

The assertion then follows from the condition (27). ∎

From Lemmas 18 and 19, we see that a partition $T \in P(m)$, $m > 0$, contributes to the limit (24) if the inequality

$$\alpha m + \beta \kappa s(T) \le c(T) \le \frac{1}{2}(m + s(T))$$

holds.

Suppose that the sequence $(a_n^{(j)})$ satisfies the boundedness condition (3) uniformly on γ and $\varphi_\gamma(a_n^{(j)}) = c^{(j)} \gamma^\beta$, $c^{(j)} \in \mathbf{R}$ for any $n, j \in \mathbf{N}$. Then for any $T = \coprod_{i=1}^s \{s_i\} \coprod T' \in P(m)$, where $T' \in P(r)$, $r = m - s$ is a sub-partition of T with no singleton, we have

$$\left| \sum_{(n_1, \ldots, n_m) \in T_N} \varphi_{\gamma(N)} \left(a_{n_1}^{(1)} \cdots a_{n_m}^{(m)} \right) \right|$$

$$\le C_m \cdots C_{m-s+1} \frac{|c^{(s_1)} \cdots c^{(s_s)}|}{N^{\alpha m}} \gamma(N)^{\beta s} c(T)! \binom{N}{c(T)} |\varphi_{\gamma(N)}(T')|$$

$$= C_m \cdots C_{m-s+1} |c^{(s_1)} \cdots c^{(s_s)}| \frac{\lambda^{\beta s} c(T)!}{N^{\alpha m + \beta \kappa s}} \binom{N}{c(T)} |\varphi_{\gamma(N)}(T')| \tag{28}$$

Here we put $\varphi_\gamma(T') = \varphi_\gamma(a_{n_1}^{(j_1)} \cdots a_{n_r}^{(j_r)})$ for $(n_1, \ldots, n_r) \in T'$. By Lemma 19 and (28), in order to obtain finite moments of all order it is sufficient to prove that for any $T \in P(m)$,

$$c(T) \le \alpha m + \beta \kappa s(T) \qquad (29)$$

holds. Only the partition that holds the equality in (29) contribute to the limit. Suppose that $m = 2p > 0$ and take a partition $T \in PP(2p)$. Then $c(T) = p$ and $s(T) = 0$. It follows from (29) that $\alpha \ge 1/2$. Suppose that a partition $T \in P(m)$ holds the equality in (29). It follows from Lemma 18 and $\alpha \ge 1/2$ that

$$0 \le (\frac{1}{2} - \alpha)m + (\frac{1}{2} - \beta\kappa)s(T) \le (\frac{1}{2} - \beta\kappa)s(T). \qquad (30)$$

Suppose that $\alpha > 1/2$, then the partition T contains a singleton, that is, $s(T) \ne 0$, otherwise the inequality (30) breaks down. Thus we have $\beta\kappa < 1/2$. Since $s(T) \le m$, we have

$$0 \le (\frac{1}{2} - \alpha)m + (\frac{1}{2} - \beta\kappa)s(T) \le (1 - \alpha - \beta\kappa)m,$$

hence

$$\alpha + \beta\kappa \le 1. \qquad (31)$$

Take a partition $T \in P(m)$ with $c(T) = s(T) = m$. It follows from (29) and (31) that $\alpha + \beta\kappa = 1$.

Suppose that $\alpha = 1/2$. Again take a partition $T \in P(m)$ with $c(T) = s(T) = m$. From (29) we have $\beta\kappa \ge 1/2$. If $\beta\kappa > 1/2$, $s(T) = 0$ holds from (30), and then we see from Lemma 18 that only pair partitions may survive in the limit.

Let us consider the converse. Suppose that $\alpha > 1/2$ and $\alpha + \beta\kappa = 1$. We see that for $T \in P(m)$,

$$\alpha m + \beta\kappa s(T) - \frac{m + s(T)}{2} = \left(\alpha - \frac{1}{2}\right)m + \left(\beta\kappa - \frac{1}{2}\right)s(T)$$
$$\ge (\alpha + \beta\kappa - 1)m = 0$$

since $s(T) \le m$ and $\beta\kappa < 1/2$. Then Lemma 18 shows the inequality (29). The equality in (29) holds if and only if $c(T) = s(T) = m$. Suppose that $\alpha = 1/2$ and $\beta\kappa > 1/2$. Combining with Lemma 18, we have the inequality (29). The equality holds only for the partitions $T \in P(m)$ with $s(T) = 0$ and $m = 2c(T)$, that is, the pair partitions. Finally, suppose that $\alpha = \beta\kappa = 1/2$, Lemma 18 shows (29) and its equality holds only when the partition T consists of pairs and singletons. In summary, we come to the following limit theorem.

Theorem 20 *Assume that sequences $(a_n^{(j)}) \subset \mathcal{A}$ are singleton independent and satisfy the boundedness condition (3) uniformly, with respect to a family of states φ_γ, $\gamma \geq 0$. Let $\gamma(N) = \lambda/N^\kappa$ and suppose that for each j there exists $c^{(j)} \in \mathbf{R}$ such that $\varphi_\gamma(a_n^{(j)}) = c^{(j)}\gamma^\beta$ for any n. Then for any m, the mixed moment (24) has non-zero finite limit if one of the followings holds:*

(i) $\alpha + \beta\kappa = 1$ *and* $\alpha > 1/2$. *In this case the partition $T \in P(m)$ may survive in the limit only if $c(T) = s(T) = m$, that is, T consists of singletons only. Then the mixed moment (24) of order m is given by*

$$\lim_{N \to \infty} \frac{1}{N^{\alpha m}} \sum_{(n_1,\ldots,n_m) \in T_N} \varphi_{\gamma(N)}(a_{n_1}^{(1)} \cdots a_{n_m}^{(m)}),$$

where $T \in P(m)$ with $c(T) = s(T) = m$, that is, the partition consists of singletons only. The moment is bounded from above by

$$C_1 \cdots C_m |c^{(1)} \cdots c^{(m)}| \lambda^{\beta m}.$$

(ii) $\alpha = 1/2$ *and* $\beta\kappa > 1/2$. *In this case the partition $T \in P(m)$ may survive in the limit only if T is a pair partition. The mixed moment (24) of order $2m$ is given by*

$$\lim_{N \to \infty} \frac{1}{N^m} \sum_{T \in PP(2m)} \sum_{(n_1,\ldots,n_{2m}) \in T_N} \varphi_{\gamma(N)}(a_{n_1}^{(1)} \cdots a_{n_{2m}}^{(2m)}).$$

(iii) $\alpha = \beta\kappa = 1/2$. *In this case the partition $T \in P(m)$ may survive in the limit only if the all blocks S_j contained in T satisfy $\#S_j \leq 2$, that is, the partition consists of pairs and singletons only. The mixed moment (24) of order m is given by*

$$\lim_{N \to \infty} \frac{1}{N^{m/2}} \sum_{s=0}^{m} \sum_{T \in PP(m;s)} \sum_{(n_1,\ldots,n_m) \in T_N} \varphi_{\gamma(N)}(a_{n_1}^{(1)} \cdots a_{n_m}^{(m)}),$$

where $PP(m;s) \subset P(m)$ stands for the set of partitions consists of pairs and s singletons.

5   A limit theorem on the Haagerup states

The notion of singleton independence is found in investigation of the Haagerup state on a free group, The Haagerup state is a non-trivial instance that fails to satisfy the singleton condition. Here we give a summary of the studies[2,3] on combinatorial aspects of the Haagerup state.

Let F be the free group on countably infinite generators $\Sigma = \{g_n \mid n \in \mathbf{Z}^\times\}$, where g_{-n} stands for g_n^{-1}. Denote by $\mathcal{A}_F$ the group $*$-algebra of F and for each $n \in \mathbf{N}$ by $\mathcal{A}_n$ the $*$-subalgebra generated by $\{g_{\pm n}\}$. For $0 \le \gamma < 1$ we denote by φ_γ the Haagerup state defined by

$$\varphi_\gamma(g) = \gamma^{|g|}, \qquad g \in F, \tag{32}$$

where $|\cdot|$ denotes the reduced length function, that is, for $g = g_{n_1}^{\varepsilon_1} \cdots g_{n_m}^{\varepsilon_m}$ with $|n_1| \ne |n_2| \ne \cdots \ne |n_m|$, its length $|g|$ is given by $|\varepsilon_1| + \cdots + |\varepsilon_m|$. We set $|e| = 0$ for the unit $e \in F$ and $0^0 = 1$. $\{g_n\}$ satisfy the singleton condition with respect to φ_γ if and only if $\gamma = 0$. In particular, the algebras $\mathcal{A}_n$ are not free independent with respect to φ_γ, $\gamma \ne 0$. However we have the following.

Proposition 21 *The sequence $\{g_n \mid n \in \mathbf{Z}^\times\}$ are singleton independent with respect to the Haagerup states φ_γ, $0 \le \gamma < 1$.*

In view of the reduction $g_n \cdot g_{-n} = e$, we attach a partition $T \in P(m)$ to each index tuple $(\alpha_1, \dots, \alpha_m)$ in a slightly different way from Notation 2: for an m-tuple $(\alpha_1, \dots, \alpha_m) \in (\mathbf{Z}^\times)^m$, the partition $T = [|\alpha_1, \dots, \alpha_m|] \in P(m)$ is defined by

$$\coprod_{i=1}^{s} \{s_i\} \coprod_{j=1}^{t} S_j,$$

where $\alpha_{s_i} \ne -\alpha_k$ for any $1 \le k \le m$ and S_j's are given by the form

$$S_j = \{t_1, \dots, t_l \mid |\alpha_{t_1}| = \cdots = |\alpha_{t_l}| \text{ and } \exists p, q \text{ such that } \alpha_{t_p} = -\alpha_{t_q}\}.$$

The factor α_{s_i} is called a *signed singleton*. A signed singleton α_{s_i} is said to be *inner* if there is a block S_j such that $\underline{s}_j < s_i < \overline{s}_j$ (see Definition 7), and is said to be *outer* if there is no such a block. $i(T)$ and $b(T)$ denote the number of inner singletons and the number of blocks which are not signed singletons, respectively. An m-tuple $(\alpha_1, \dots, \alpha_m)$ is said to be *separable at k* if there is no block S_j such that $\underline{s}_j \le k < \overline{s}_j$. Note that the existence of an outer singleton α_s implies the separability at $s - 1$ and s.

Definition 22 Let $NCPI(m; s) \subset P(m)$ be the set of partitions which consist of non-crossing pair partitions and s inner singletons, that is, an element $T \in NCPI(m; s)$ is given by

$$T = \coprod_{i=1}^{s} \{s_i\} \coprod_{j=1}^{t} \{p_j, q_j\},$$

where s_i's are inner singletons and $\coprod_{j=1}^{t} \{p_j, q_j\}$ forms a non-crossing pair partition.

64

Theorem 23 [2,3] *For $T \in P(m)$ with s inner singletons, we have*

$$\lim_{N \to \infty} \sum_{(\alpha_1,\ldots,\alpha_m) \in T_N} \frac{1}{(\sqrt{N})^m} \varphi_{\lambda/\sqrt{N}}(\tilde{g}_{\alpha_1} \cdots \tilde{g}_{\alpha_m})$$

$$= \begin{cases} (-\lambda)^s, & if \ T \in NCPI(m;s), \\[2ex] 0, & otherwise. \end{cases} \tag{33}$$

Here we put $\tilde{g}_\alpha = g_\alpha - \gamma$.

Hence for $a_N^\pm := (\tilde{g}_{\pm 1} + \cdots + \tilde{g}_{\pm N})/\sqrt{N}$, the mixed momenta $\varphi_{\lambda/\sqrt{N}}(a_N^{\varepsilon_1} \cdots a_N^{\varepsilon_m})$ with $\varepsilon = (\varepsilon_1,\ldots,\varepsilon_m)$, $\varepsilon_i = \pm$, in the limit is given by

$$\sum_{s=0}^{m-2} (-\lambda)^s \cdot \#NCPI(m;s;\varepsilon).$$

Here $NCPI(m;s;\varepsilon)$ denotes the subset of $NCPI(m;s)$, that consists of m-tuple $(\alpha_1,\ldots,\alpha_m)$ where the sign of each α_i equals to ε_i.

In view of Theorem 23, we construct a representation of the limit process for the Haagerup states on a GNS space. Let

$$\Gamma(\mathbf{C})_\nu = \mathbf{C} \oplus \bigoplus_{n=1}^{\infty} \mathbf{C}^{\otimes n} \quad \left(= \bigoplus_{n=0}^{\infty} \mathbf{C}\right), \qquad \nu = L, R$$

denote two copies of the full Fock spaces on $\mathbf{C}$ with free creations a_ν^+ and free annihilation a_ν. Let $\mathcal{H} = \bigoplus_{m,n=0}^{\infty} \mathcal{H}_{m,n}$ be the free product $\Gamma(\mathbf{C})_L * \Gamma(\mathbf{C})_R$, that is, its (m,n)-particle space $\mathcal{H}_{m,n}$ is the complex linear span of the vectors $\{|\nu_1,\ldots,\nu_{m+n}\rangle \mid \#\{j \mid \nu_j = L\} = m, \ \#\{j \mid \nu_j = R\} = n\}$ equipped with a scalar product given by

$$\langle \nu_1,\ldots,\nu_k | \nu_1',\ldots,\nu_l' \rangle_{\mathcal{H}} = \delta_{(\nu_1,\ldots,\nu_k),(\nu_1',\ldots,\nu_l')}.$$

The creation operators

$$L^+ := a_L^+ * 1 : \mathcal{H}_{m,n} \to \mathcal{H}_{m+1,n} \ ; \quad R^+ := 1 * a_R^+ : \mathcal{H}_{m,n} \to \mathcal{H}_{m,n+1}$$

are given respectively by

$$L^+|\nu_1,\ldots,\nu_k\rangle = |L,\nu_1,\ldots,\nu_k\rangle, \qquad R^+|\nu_1,\ldots,\nu_k\rangle = |R,\nu_1,\ldots,\nu_k\rangle,$$

and the annihilation operators are given by the duals $L = (L^+)^*$ and $R = (R^+)^*$. Let $P : \mathcal{H} \to \mathcal{H}$ be the orthogonal projection onto $\mathcal{H}_{0,0}^\perp$, the subspace orthogonal to $\mathcal{H}_{0,0} = \mathbf{C}|0\rangle$, where $|0\rangle$ denotes the vacuum vector. Put

$$A_\lambda^- = L^+ + R - \lambda P, \qquad A_\lambda^+ = L + R^+ - \lambda P,$$

where $\lambda \geq 0$ is a constant.

Theorem 24 [2,3] *The limit process for $(a_N^+, a_N^-, \varphi_{\lambda/\sqrt{N}})$ is represented on $\mathcal{H}$. That is, all its correlations in the limit are given by*

$$\lim_{N \to \infty} \varphi_{\lambda/\sqrt{N}}(a^{\varepsilon_1} \cdots a^{\varepsilon_m}) = \langle \Phi, A_\lambda^{\varepsilon_1} \cdots A_\lambda^{\varepsilon_m} \Phi \rangle_{\mathcal{H}}.$$

Remark. We have seen that the field operator $a_N^+ + a_N^-$ converges in the sense of correlations to

$$A_\lambda^+ + A_\lambda^- = (L^+ + R^+) + (L^- + R^-) - 2\lambda P.$$

This means that the field operator is decomposed into a sum of 'creation' $L^+ + R^+$, 'annihilation' $L^- + R^-$ and a function of the number operator, $-\lambda P$. This decomposition leads us naturally to an idea of 'quantum decomposition' introduced in the next section.

6 Quantum decomposition in a discrete group

We here introduce a new idea of 'quantum' decomposition of a graph Laplacian on a Cayley graph associated with a discrete group. To make the point clear, we focus on discrete groups satisfying condition (A2) below. More general situations are discussed in section 8.3 and 8.4.

Let G be a discrete group equipped with a length function $|\cdot| : G \to \mathbf{N}_0 :=$ $\mathbf{N} \cup \{0\}$ and a set of generators $\Sigma := \{g_\alpha \mid \alpha \in \mathbf{Z}^\times\}$. The length of the unit $e \in G$ is 0. Throughout this section we adopt the convention $\{g_{-\alpha} = g_\alpha^{-1}\}$ and assume $g_{-\alpha} \neq g_\alpha$ for simplicity. We say that Σ is *compatible* with respect to the length function $|\cdot|$ if

(A1) $|g_\alpha| = 1$ for any $g_\alpha \in \Sigma$,

(A2) $|g_\alpha \cdot g| = |g| \pm 1$ for any $g_\alpha \in \Sigma$ and $g \in G$.

The property (A2) leads us to the following idea: we take the left regular representation $\pi : G \to \mathbf{B}(l^2(G))$. Let $\pi(g_\alpha^\pm) \in \mathbf{B}(l^2(G))$ be bounded operators defined by

$$\pi(g_\alpha^+)\delta_g = \begin{cases} \delta_{g_\alpha g}, & \text{if } |g_\alpha g| = |g| + 1, \\ 0, & \text{otherwise}, \end{cases} \qquad \pi(g_\alpha^-)\delta_g = \begin{cases} \delta_{g_\alpha g}, & \text{if } |g_\alpha g| = |g| - 1, \\ 0, & \text{otherwise}. \end{cases}$$

Here δ_g denotes the characteristic function of a singlet $\{g\}$. Then we have a 'quantum' decomposition of g_α;

$$\pi(g_\alpha) = \pi(g_\alpha^+) + \pi(g_\alpha^-). \tag{34}$$

We also see that $\pi(g_\alpha^+)^* = \pi(g_{-\alpha}^-)$ and $\|\pi(g_\alpha^\pm)f\|_{l^2} \leq \|f\|_{l^2}$ for $f \in l^2(G)$.

A family of subsets $\Sigma^{(N)} := \{g_\alpha \mid |\alpha| = 1, \ldots, N\} \subset \Sigma$ gives a filtration of G: $G^{(1)} \subset G^{(2)} \subset \cdots \subset G$, where $G^{(N)}$ is the subgroup generated by $\Sigma^{(N)}$. For each $g \in G^{(N)}$ put

$$\omega_+^{(N)}(g) := \{(g_\alpha, x) \in \Sigma^{(N)} \times G^{(N)} \mid \pi(g_\alpha^+)\delta_x = \delta_g\},$$

$$\omega_-^{(N)}(g) := \{(g_\alpha, x) \in \Sigma^{(N)} \times G^{(N)} \mid \pi(g_\alpha^-)\delta_x = \delta_g\}.$$

We note that $\pi(g_\alpha^\pm)\delta_x = \delta_g$ implies $|g| = |x| \pm 1$ and $\#\omega_+^{(N)}(g) + \#\omega_-^{(N)}(g) = 2N$ by assumption (A2). We also note that for the unit $e \in G$ we have $\omega_+^{(N)}(e) = \emptyset$ and $\omega_-^{(N)}(e) = \Sigma^{(N)}$ by definition. With these notations we assume

(A3) for each $n \in \mathbf{N}$, there exist $\omega_n \in \mathbf{N}$ and $C_n > 0$ such that

$$\#\{g \in G^{(N)} \mid |g| = n \text{ and } \#\omega_+^{(N)}(g) \neq \omega_n\} \leq C_n(2N)^{n-1},$$

(A4) for each $n \in \mathbf{N}$,

$$\sup_{N \in \mathbf{N}} \sup_{g \in G^{(N)}, |g|=n} \#\omega_+^{(N)}(g) =: W_n < \infty,$$

and $\limsup_{n \to \infty} W_n^{1/n} < \infty$.

(A5) if $|g| = n$ and $g \in G^{(N)}$, there exists an n-tuple $(g_{\alpha_1}, \ldots, g_{\alpha_n}) \in (\Sigma^{(N)})^n$ such that

$$\delta_g = \pi(g_{\alpha_1}^+) \cdots \pi(g_{\alpha_n}^+)\delta_e.$$

Canonical examples associated with a free group and a free abelian group are again observed in section 8.1.

7   Interacting Fock spaces associated with a discrete group

For each subgroup $G^{(N)}$, we construct a one-mode interacting Fock space as follows. See Hashimoto[10] for proofs. The vacuum vector is defined by $|0\rangle := \delta_e$. A vector $v \in l^2(G^{(N)})$ is called n-*homogeneous* if it has a form of

$$v = \sum_{g \in G^{(N)}, |g|=n} v(g)\pi(g)|0\rangle, \qquad v(g) \in \mathbf{C}. \tag{35}$$

Definition 25 (1) Let $\Gamma_n^{(N)}$ be a one-dimensional complex vector space equipped with a pre-scalar product

$$(z|w)_n^{(N)} := \bar{z}w\langle\Phi(n)|\Phi(n)\rangle^{(N)}, \qquad z, w \in \mathbf{C},$$

where $|\Phi(n)\rangle^{(N)}$ stands for a *number vector* defined by

$$|\Phi(n)\rangle^{(N)} := \frac{(\omega_n)!}{(\sqrt{2N})^n} \sum_{g \in G^{(N)},\ |g|=n} \pi(g)|0\rangle,$$

with $(\omega_n)! = \omega_1 \omega_2 \cdots \omega_n$. The completion of the orthogonal sum $\oplus_n \{\Gamma_n^{(N)}, (\cdot|\cdot)_n^{(N)}\}$ is denoted by $\Gamma(G^{(N)})$, which we call the *interacting Fock space associated with $G^{(N)}$*.

(2) The operators

$$a_N^+ := \left(\pi(g_{-N}^+) + \cdots + \pi(g_{-1}^+) + \pi(g_1^+) + \cdots + \pi(g_N^+)\right)/\sqrt{N}$$

$$a_N^- := \left(\pi(g_{-N}^-) + \cdots + \pi(g_{-1}^-) + \pi(g_1^-) + \cdots + \pi(g_N^-)\right)/\sqrt{N}$$

are called *asymptotic creation* and *annihilation operators* respectively.

(3) For $\lambda \in \mathbf{C}$ we define

$$|\mathcal{E}(\lambda)\rangle^{(N)} := \sum_{n=0}^{\infty} \frac{\lambda^n}{(\omega_n)!}|\Phi(n)\rangle^{(N)},$$

whenever $\langle \mathcal{E}(\lambda)|\mathcal{E}(\lambda)\rangle = \sum_{n=0}^{\infty} |\lambda|^{2n}/(\omega_n)!$ converges. We call $|\mathcal{E}(\lambda)\rangle^{(N)}$ an *asymptotic coherent vector*.

Lemma 26 [10]

(1) *For any $n \geq 0$,*

$$a_N^+|\Phi(n)\rangle^{(N)} = |n+1\rangle^{(N)} + v_{n+1}\left(\frac{1}{\sqrt{N}}\right),$$

$$a_N^-|\Phi(n+1)\rangle^{(N)} = \omega_{n+1}|n\rangle^{(N)} + v_n\left(\frac{1}{N}\right), \quad and \quad a_N^-|\Phi(0)\rangle = 0,$$

where $v_n(N^m)$ is an n-homogeneous vector such that $\|v_n(N^m)\|_{l^2} = O(N^m)$.

(2) *For each $n \geq 0$ and $N \geq 1$ it holds that*

$$\langle \Phi(n)|\Phi(n)\rangle^{(N)} = (\omega_n)! + O\left(\frac{1}{N}\right).$$

(3) *For $\lambda \in \mathbf{C}$ in a neighbourhood of 0 we have*

$$a_N^-|\mathcal{E}(\lambda)\rangle^{(N)} = \lambda|\mathcal{E}(\lambda)\rangle^{(N)} + v\left(\frac{1}{N}\right).$$

(4) *Let $v \in l^2(G^{(N)})$ be an n-homogeneous vector. Then, $a_N^+ v$ and $a_N^- v$ are respectively $(n+1)$- and $(n-1)$-homogeneous vectors with the norm estimates:*

$$\|a_N^+ v\|_{l^2}^2 \leq 2NW_n\|v\|_{l^2}^2, \qquad \|a_N^- v\|_{l^2}^2 \leq 2NW_n\|v\|_{l^2}^2.$$

By virtue of lemma 26 we associate a *one-mode interacting Fock space* $\Gamma(G^{(N)})$ with parameters $\{\lambda_0 := 1, \lambda_n := (\omega_n)!\}$ for each $G^{(N)}$. (See Accardi-Bożejko[1] for definitions.)

Next, we construct an interacting Fock space for the limit of a_N^+ and a_N^- as $N \to \infty$. Let $\Gamma(G)$ denote the one-mode interacting Fock space with parameters $\{\lambda_0 := 1, \lambda_n := (\omega_n)!\}$. By definition $\Gamma(G)$ is the completion of the orthogonal sum of one-dimensional space $\Gamma_n := \mathbf{C}|\Phi(n)\rangle$ equipped with a pre-scalar product

$$(z|w)_n := \bar{z}w\langle\Phi(n)|\Phi(n)\rangle, \qquad z, w \in \mathbf{C},$$

where $|\Phi(n)\rangle$ stands for a number vector with norm $\sqrt{\lambda_n} = \sqrt{(\omega_n)!}$. The creation a^+ and annihilation a^- are uniquely determined by

$$a^+|\Phi(n)\rangle = |\Phi(n+1)\rangle, \qquad a^-|\Phi(n+1)\rangle = \omega_{n+1}|\Phi(n)\rangle, \qquad a^-|\Phi(0)\rangle = 0.$$

For $\lambda \in \mathbf{C}$, we define a coherent vector

$$|\mathcal{E}(\lambda)\rangle := \sum_{n=0}^{\infty} \frac{\lambda^n}{(\omega_n)!}|\Phi(n)\rangle \tag{36}$$

whenever $\langle\mathcal{E}(\lambda)|\mathcal{E}(\lambda)\rangle = \sum_{n=0}^{\infty}|\lambda|^{2n}/(\omega_n)!$ converges. By definition, it holds that $a^-|\mathcal{E}(\lambda)\rangle = \lambda|\mathcal{E}(\lambda)\rangle$. For $u = \sum u_n|\Phi(n)\rangle \in \Gamma(G)$, we put

$$u^{(N)} := \sum u_n|\Phi(n)\rangle^{(N)} \in \Gamma(G^{(N)}). \tag{37}$$

Theorem 27 [10]*Let $m \geq 1$ and $\epsilon_1, \ldots, \epsilon_m \in \{\pm\}$. Then, for any $u \in \Gamma(G)$ and $n \in \mathbf{N}_0$ we have*

$$\lim_{N\to\infty} \langle u^{(N)}|a_N^{\epsilon_1}\cdots a_N^{\epsilon_m}|\Phi(n)\rangle_{\Gamma(G^{(N)})}^{(N)} = \langle u|a^{\epsilon_1}\cdots a^{\epsilon_m}|\Phi(n)\rangle_{\Gamma(G)}.$$

8 Applications to limit theorems in algebraic probability theory

8.1 Central limit theorems on the vacuum state

Since Theorem 27 is nothing but a general form of limit theorems in algebraic probability theory, it particularly leads to the algebraic central limit theorem with respect to the vacuum state $\langle 0| \cdot |0\rangle$:

$$\lim_{N \to \infty} \langle 0| \Big(\frac{a_N^+ + a_N^-}{\sqrt{2N}} \Big)^m |0\rangle_{\Gamma(G^{(N)})}^{(N)} = \langle \Phi(0)|(a^+ + a^-)^m|\Phi(0)\rangle_{\Gamma(G)}. \qquad (38)$$

Let A (resp. F) be a free abelian group (resp. a free group) generated by $\Sigma := \{g_\alpha\}$, where $g_\alpha^{-1} = g_{-\alpha}$ according to our convention. Each element $g \in A$ (resp. $g \in F$) has a canonical expression in terms of Σ:

$$g = g_{\alpha_1}^{\epsilon_1} \cdots g_{\alpha_m}^{\epsilon_m},$$

where $0 < \alpha_1 < \alpha_2 < \cdots < \alpha_m$, $\epsilon_i \in \mathbf{Z}$ (resp. $|\alpha_1| \neq |\alpha_2| \neq \cdots \neq |\alpha_m|$, $\epsilon_i \in \mathbf{Z}$), and a reduced length function is defined by

$$|g| = |\epsilon_1| + \cdots + |\epsilon_m|.$$

Conditions (A1)(A2)(A5) are obvious. As for (A3)(A4), we see for $g \in A$ (resp. $g \in F$) with $|g| = n \geq 1$,

$$\#\omega_+^{(N)}(g) \leq n, \quad \#\{g \in A \mid |g| = n, \#\omega_+^{(N)}(g) < n\} \leq (2N)^{n-1},$$

$$2N - n \leq \#\omega_-^{(N)}(g) \leq 2N - 1,$$

$$(\text{ resp. } \#\omega_+^{(N)}(g) = 1, \ \#\omega_-^{(N)}(g) = 2N - 1).$$

Hence we have $\omega_n = n$ (resp. $\omega_n = 1$), which implies that the interacting Fock space in the limit is the one-mode boson (resp. free) Fock space.

8.2 Limit theorems for Haagerup states

In this paragraph, we again observe the limit process associated with the Haagerup state (32). Contrast to Theorem 24, we here give a fully quantum representation of the limit process. The result in Hashimoto[8] was thought as an analogy of a central limit theorem for a vacuum state, though the limit distribution had peculiar properties. In fact, the limit distribution coincides with the free Poisson law[23] up to translation. It is noticeable that limit behaviors of the field operators under a vacuum state are completely described by pair-partition statistics, as is seen in Lemma 4, while the Poisson laws are induced from all set partition statistics[13,21,20]. Proposition 28 implies that the Haagerup states give rise to a transform of set partition statistics through a coherent state expression of the Haagerup functions. See Hashimoto[10] for proofs.

We note that for any $\lambda \in \mathbf{C}$ with $|\lambda| < 1$ and $g \in F^{(N)}$, we have

$$\varphi_{\frac{\lambda}{\sqrt{2N}}}(g) = \langle \mathcal{E}(\lambda)|\pi(g)|\Phi(0)\rangle_{\Gamma(F^{(N)})}^{(N)}.$$

It follows from Theorem 27 that the moments of the field operator in the limit are given by

$$\lim_{N \to \infty} \varphi_{\frac{\lambda}{\sqrt{2N}}} \left(\left(\frac{a_N^+ + a_N^-}{\sqrt{2N}} \right)^m \right) = \langle \mathcal{E}(\lambda) | (a^+ + a^-)^m | \Phi(0) \rangle_{\Gamma(F)}. \qquad (39)$$

Note that $\langle \Phi(n) | a^+ a^- | \Phi(n) \rangle = 1 = \langle \Phi(0) | P_0 | \Phi(0) \rangle$ holds for $n \geq 1$, where P_0 denotes the vacuum projection. Putting $\phi_0(A) = \langle \Phi(0) | A | \Phi(0) \rangle$, we have a recurrence formula

$$\langle \mathcal{E}(\lambda) | (a^+ + a^-)^m | \Phi(0) \rangle = \phi_0 \big((a^+ + a^-)^m \big) \qquad (40)$$
$$+ \sum_{l=0}^{\lfloor (m-1)/2 \rfloor} \phi_0 \big((a^+ + a^-)^{2l} \big) \phi_0(\lambda P_0) \cdot \langle \mathcal{E}(\lambda) | (a^+ + a^-)^{m-2l-1} | \Phi(0) \rangle,$$

while the expansion of $(a^+ + a^- + \lambda P_0)^m$ gives

$$\phi \big((a^+ + a^- + \lambda P_0)^m \big) = \phi \big((a^+ + a^-)^m \big)$$
$$+ \sum_{l=0}^{\lfloor (m-1)/2 \rfloor} \phi \big((a^+ + a^-)^{2l} \big) \phi(\lambda P_0) \phi \big((a^+ + a^- + \lambda P_0)^{m-2l-1} \big).$$

By induction we then obtain the following.

Proposition 28 [10] *In the free Fock space* $\{\Gamma(F), \{\lambda_n \equiv 1\}\}$*, the identity*

$$\langle \mathcal{E}(\lambda) | (a^+ + a^-)^m | \Phi(0) \rangle_{\Gamma(F)} = \langle \Phi(0) | (a^+ + a^- + \lambda P_0)^m | \Phi(0) \rangle_{\Gamma(F)}$$

holds, where P_0 is the vacuum projection.

Remarks.

(1) Since in the free Fock space the vacuum projection is given by $P_0 = I - a^+ a^-$, the random variable $a^+ + a^- + \lambda P_0$ is nothing but the free Poisson one [21], that is,

$$a^+ + a^- + \lambda P_0 = \lambda + \frac{1}{\lambda} - \left(\sqrt{\lambda} a^+ - \frac{1}{\sqrt{\lambda}} \right) \left(\sqrt{\lambda} a^- - \frac{1}{\sqrt{\lambda}} \right).$$

(2) Proposition 28 describes a *Gaussian-Poisson transform* in terms of coherent vectors, while Oravecz[19] recently introduced a Gaussian-Poisson transform called the *$\mathcal{F}$-transform* for an arbitrary symmetric measure from the viewpoint of orthogonal polynomials.

(3) With the recurrence formula (40) one obtains a functional identity for the moment generating functions. Put $F(t) = 1 + \sum_{n=1}^{\infty} F_m t^m$ with $F_m = \langle \mathcal{E}(\lambda) | (a^+ + a^-)^m | \Phi(0) \rangle$ and $C(t) = (1 - \sqrt{1 - 4t^2})/2t^2$, which is

the moment generating function of the normalized semi-circle law. Then we have the identity $F(t) = C(t) + \lambda t C(t) F(t)$, from which the limit distribution is computed easily:

$$d\mu(x) = \max\left\{0, \frac{1}{2}\left(1 - \frac{1}{\lambda^2}\right)\right\} \delta_{\lambda + \frac{1}{\lambda}}(x) + \frac{1}{2\pi} \chi_{[-2,2]} \frac{\sqrt{4 - x^2}}{\lambda^2 + 1 - \lambda x} dx.$$

8.3   Multiplication of free elements

Proposition 28 implies that under the Haagerup state, $S_N/\sqrt{2N} = \pi(g_1) + \pi(g_1^{-1}) + \cdots + \pi(g_N) + \pi(g_N^{-1})$ is decomposed into a creation a^+, an annihilation a^- and conservation $a^+ a^-$ in the limit. In this subsection we present a central limit theorem associated with a multiplication of free elements, and again we obtain the same decomposition as Proposition 28.

Let F be a free product of $\mathbf{Z}/2\mathbf{Z}$, $F = *_{i=1}^{\infty} \mathbf{Z}/2\mathbf{Z}\sigma_i$, where σ_i's are generator of order 2. Note that σ_i's are free from each other with respect to the vacuum state $\langle 0| \cdot |0\rangle$ in the sense of Voiculescu [23]. Let us consider a set of products of free elements $\Sigma_2 := \{w_{ij} := \sigma_i \sigma_j \, (i \neq j)\}$, which are not free from each other, and a subgroup $F_2 \subset F$ generated by Σ_2. We use a new length function $|\cdot|_2 : F_2 \to \mathbf{N}_0$ with respect to Σ_2 different from the one introduced in Section 8.1. Since any product $w_{i_1 j_1} \cdots w_{i_m j_m}$ has a reduced expression in terms of σ_i's,

$$w_{i_1 j_1} \cdots w_{i_m j_m} = \sigma_{k_1} \cdots \sigma_{k_l},$$

where $k_1 \neq k_2 \neq \cdots \neq k_l$, its length is defined by

$$|w_{i_1 j_1} \cdots w_{i_m j_m}|_2 := l/2. \tag{41}$$

Then we see for any $g = \sigma_{k_1} \cdots \sigma_{k_m} \in F_2$ and w_{ij}, the following three cases occur:

$$|w_{ij} g|_2 = \begin{cases} |g|_2 + 1, & \text{if } j \neq k_1, \\ |g|_2 - 1, & \text{if } (i,j) = (k_2, k_1), \\ |g|_2, & \text{if } j = k_1 \text{ and } i \neq k_2. \end{cases} \tag{42}$$

It follows from $|w_{ij}|_2 = 1$ and (42) that $|g|_2 \in \mathbf{N}_0$ for any $g \in F_2$.

The existence of the third case is a noticeable difference from our discussions before. Concerning the new case, we extend our argument. The condition (A2) in Section 6 is replaced to

(A2)' $|g_\alpha \cdot g| = |g|$ or $|g| \pm 1$ for any $g_\alpha \in \Sigma$ and $g \in G$.

In accordance with (A2)' we have a 'quantum' decomposition of g_α:

$$g_\alpha = g_\alpha^+ + g_\alpha^- + g_\alpha^\circ, \tag{43}$$

where $g_\alpha^\pm$ are defined by (34) and g_α° is given by

$$\pi(g_\alpha^\circ)\delta_g = \begin{cases} \delta_{g_\alpha g}, & \text{if } |g_\alpha g| = |g|, \\ 0, & \text{otherwise.} \end{cases}$$

Let us return to the example (F_2, Σ_2). We put $\Sigma_2^{(N)} := \{w_{ij} \mid 1 \le i \ne j \le N\}$. $F_2^{(N)}$ denotes the subgroup generated by $\Sigma_2^{(N)}$. For $g \in F_2^{(N)}$ we define $\omega_\pm^{(N)}(g)$ by (35) where we take $G^{(N)}$ for $F_2^{(N)}$, and $\omega_\circ^{(N)}(g)$ by

$$\omega_\circ^{(N)}(g) := \{(w_{ij}, x) \in \Sigma_2^{(N)} \times F_2^{(N)} \mid \pi(g_\alpha^\circ)\delta_x = \delta_g\}. \tag{44}$$

By definition, we have $\omega_+^{(N)}(e) = \omega_\circ^{(N)}(e) = \emptyset$ and $\omega_-^{(N)}(e) = \Sigma_2^{(N)}$ for the unit $e \in F_2$. It is easy to see that for $g \in F_2^{(N)}$ with $|g|_2 = n \ge 1$ and for large N,

$$\#\omega_+^{(N)}(g) = 1, \quad \#\omega_-^{(N)}(g) = (N-1)^2 \quad \text{and} \quad \#\omega_\circ^{(N)}(g) = N - 2.$$

Also note that $\#\omega_+^{(N)}(g) + \#\omega_-^{(N)}(g) + \#\omega_\circ^{(N)}(g) = \#\Sigma_2^{(N)} = N(N-1)$. According to Section 7, we define a number vector

$$|\Phi(n)\rangle^{(N)} := \left(\frac{1}{\sqrt{\#\Sigma_2^{(N)}}}\right)^n \sum_{g \in F^{(N)}(2), |g|_2 = n} \pi(g)|0\rangle,$$

and define canonical operators corresponding to the decomposition (43),

$$a_N^\epsilon := \frac{1}{\sqrt{\#\Sigma_2^{(N)}}} \sum_{1 \le i \ne j \le N} \pi(w_{ij}^\epsilon),$$

where $\epsilon = \pm, \circ$. Then by similar arguments in Section 7, we have

$$a_N^+|\Phi(n)\rangle^{(N)} = |\Phi(n+1)\rangle^{(N)} + v_{n+1}\left(\frac{1}{N}\right),$$

$$a_N^-|\Phi(n+1)\rangle^{(N)} = |\Phi(n)\rangle^{(N)} + v_n\left(\frac{1}{N^2}\right),$$

$$a_N^\circ|\Phi(n)\rangle^{(N)} = |\Phi(n)\rangle^{(N)} + v_n\left(\frac{1}{N}\right), \quad \text{for } n \ge 1,$$

$$a_N^-|0\rangle^{(N)} = a_N^\circ|0\rangle^{(N)} = 0, \tag{45}$$

where v_n is a homogeneous vector defined in Section 7, and then we construct an asymptotic one-mode Fock space $\Gamma(F_2^{(N)})$ associated with $F_2^{(N)}$.

Let $\Gamma(F)$ be the one-mode free Fock space, a^+, a^- and $a^\circ := P_0$ be the creation, the annihilation and the vacuum projection respectively, used in the previous subsection. It follows from (45) that we have a 'quantum' central limit theorem associated with $(F_2^{(N)}, \Sigma_2^{(N)})$ as follows.

Theorem 29 *Let $m \geq 1$ and $\epsilon_1, \ldots, \epsilon_m \in \{\pm, \circ\}$. Then for any $u \in \Gamma(F)$ and $n \in \mathbf{N}_0$, we have*

$$\lim_{N \to \infty} \langle u^{(N)} | \left(\frac{a_N^{\epsilon_1}}{\sqrt{\#\Sigma_2^{(N)}}} \right) \cdots \left(\frac{a_N^{\epsilon_m}}{\sqrt{\#\Sigma_2^{(N)}}} \right) |\Phi(n)\rangle_{\Gamma(F_2^{(N)})}^{(N)}$$
$$= \langle u | a^{\epsilon_1} \cdots a^{\epsilon_m} |\Phi(n)\rangle_{\Gamma(F)},$$

where we use the notation (37). Particularly we obtain a classical central limit theorem,

$$\lim_{N \to \infty} \langle 0 | \left(\frac{1}{\sqrt{N(N-1)}} \sum_{1 \leq i \neq j \leq N} \pi(w_{ij}) \right)^m |0\rangle_{\Gamma(F_2^{(N)})}^{(N)}$$
$$= \langle 0 | (a^+ + a^- + a^\circ)^m |0\rangle_{\Gamma(F)}.$$

8.4   Anti-commutation of semi-circular elements

Throughout this paragraph, we use the notations in Section 8.3. By taking subsets of $\Sigma_2^{(N)}$, we obtain various kinds of 'quantum' decomposition. Here we observe one of the decompositions, which leads us an analysis of anti-commutations of semi-circular elements.

Let us take a subset $\Sigma_{2,\gamma}^{(N)} \subset \Sigma_2^{(N)}$ for $0 < \gamma < 1$,

$$\Sigma_{2,\gamma}^{(N)} := \{ w_{ij} \in \Sigma_2^{(N)} \mid 1 \leq i \leq \max\{1, \gamma N\} < j \leq N$$
$$\text{or } 1 \leq j \leq \max\{1, \gamma N\} < i \leq N \}.$$

$F_{2,\gamma}^{(N)}$ denote the subgroup of $F_2^{(N)}$ generated by $\Sigma_{2,\gamma}^{(N)}$. Since (43) holds, we again obtain a decomposition

$$w_{ij} = w_{ij}^+ + w_{ij}^- + w_{ij}^\circ. \tag{46}$$

According to the decomposition, we define operators

$$a_N^\epsilon := \frac{1}{\sqrt{v}} \sum_{w_{ij} \in F_{2,\gamma}^{(N)}} \pi(w_{ij}^\epsilon), \tag{47}$$

where $\epsilon = \pm, \circ$ and $v := \#\Sigma_{2,\gamma}^{(N)} \sim 2\gamma(1-\gamma)N^2$. The purpose of this paragraph is to establish a limit theorem on the operators (47) for a constant γ. General cases are discussed in Hashimoto[9].

It is easy to see that for any $g \in F_{2,\gamma}^{(N)}$, we have

$$2(\gamma N - 1)((1 - \gamma)N - 1) \leq \#\omega_-^{(N)}(g) \leq 2\gamma(1-\gamma)N^2. \tag{48}$$

74

However, to determine $\#\omega_+^{(N)}(g)$ and $\#\omega_\circ^{(N)}(g)$, we need a delicate argument as follows. Let $I := [1, \gamma N] \cap \mathbf{N}$ and $J := (\gamma N, N] \cap \mathbf{N}$. For each $n > 0$, let us decompose the set $\mathcal{V}_n$ of elements $g \in F_{2,\gamma}^{(N)}$ with $|g|_2 = n$ into 2^{2n} disjoint subsets,

$$\mathcal{V}_n := \{g \in F_{2,\gamma}^{(N)} \mid |g|_2 = n\} = \coprod_{\eta_1, \ldots, \eta_{2n} \in \{I, J\}} \Delta(\eta_1, \ldots, \eta_{2n})$$

where $\Delta(\eta_1, \ldots, \eta_{2n})$'s are given by

$$\Delta(\eta_1, \ldots, \eta_{2n}) := \{\sigma_{i_1} \cdots \sigma_{i_{2n}} \in \mathcal{V}_n \mid i_1 \in \eta_1, \ldots, i_{2n} \in \eta_{2n}\}.$$

With the notations above, we see

$$\#\omega_+^{(N)}(g) = 1, \text{ for } g \in \Delta(I, J, \eta_3, \ldots, \eta_n) \cup \Delta(J, I, \eta_3, \ldots, \eta_{2n}),$$

$$\#\omega_+^{(N)}(g) = 0, \text{ for } g \in \Delta(I, I, \eta_3, \ldots, \eta_n) \cup \Delta(J, J, \eta_3, \ldots, \eta_{2n}),$$

$$(1 - \gamma)N - 1 \le \#\omega_\circ^{(N)}(g) \le (1 - \gamma)N, \text{ for } g \in \Delta(I, \eta_2, \ldots, \eta_{2n}),$$

$$\gamma N - 1 \le \#\omega_\circ^{(N)}(g) \le \gamma N, \text{ for } g \in \Delta(J, \eta_2, \ldots, \eta_{2n}). \tag{49}$$

According to the decomposition of $\mathcal{V}_n$, the number vector

$$|\Phi(n)\rangle^{(N)} := \left(\frac{1}{\sqrt{v}}\right)^n \sum_{g \in \mathcal{V}_n} \pi(g)|0\rangle$$

has an orthogonal decomposition into 2^{2n} homogeneous vectors

$$|\Phi(n)\rangle^{(N)} = \sum_{\eta_1, \ldots, \eta_{2n} \in \{I, J\}} |\eta_1, \ldots, \eta_{2n}\rangle^{(N)},$$

where $|\eta_1, \ldots, \eta_{2n}\rangle^{(N)} := \sum_{g \in \Delta(\eta_1, \ldots, \eta_{2n})} \pi(g)|0\rangle / \sqrt{v}$. Then we define a Fock space $\Gamma(F_{2,\gamma}^{(N)}) := \oplus_{\eta_1, \ldots, \eta_{2n} \in \{I, J\}} |\{\eta_1, \ldots, \eta_{2n}\}^{(N)}$. It follows from (48) and (49), we have

$$a_N^+(|\eta_1, \ldots, \eta_{2n}\rangle^{(N)}) = |I, J, \eta_1, \ldots, \eta_{2n}\rangle^{(N)} + |J, I, \eta_1, \ldots, \eta_{2n}\rangle^{(N)}$$

$$+ v_n\left(\frac{1}{\sqrt{v}}\right),$$

$$a_N^-|I, J, \eta_3, \ldots, \eta_{2n}\rangle^{(N)} = a_N^-|J, I, \eta_3, \ldots, \eta_{2n}\rangle^{(N)}$$

$$= \frac{1}{2}|\eta_3, \ldots, \eta_{2n}\rangle^{(N)} + v_n\left(\frac{1}{v}\right),$$

$$a_N^-|I, I, \eta_3, \ldots, \eta_{2n}\rangle^{(N)} = a_N^-|J, J, \eta_3, \ldots, \eta_{2n}\rangle^{(N)} = 0,$$

$$a_N^\circ|I, J, \eta_3, \ldots, \eta_{2n}\rangle^{(N)} = \sqrt{\frac{\gamma}{2(1 - \gamma)}}|J, J, \eta_3, \ldots, \eta_{2n}\rangle^{(N)} + v_n\left(\frac{1}{\sqrt{v}}\right),$$

$$a_N^\circ |I, I, \eta_3, \ldots, \eta_{2n}\rangle^{(N)} = \sqrt{\frac{\gamma}{2(1-\gamma)}} |J, I, \eta_3, \ldots, \eta_{2n}\rangle^{(N)} + v_n\left(\frac{1}{\sqrt{v}}\right),$$

$$a_N^\circ |J, I, \eta_3, \ldots, \eta_{2n}\rangle^{(N)} = \sqrt{\frac{1-\gamma}{2\gamma}} |I, I, \eta_3, \ldots, \eta_{2n}\rangle^{(N)} + v_n\left(\frac{1}{\sqrt{v}}\right),$$

$$a_N^\circ |J, J, \eta_3, \ldots, \eta_{2n}\rangle^{(N)} = \sqrt{\frac{1-\gamma}{2\gamma}} |I, J, \eta_3, \ldots, \eta_{2n}\rangle^{(N)} + v_n\left(\frac{1}{\sqrt{v}}\right). \quad (50)$$

In view of the relations (50), we prepare a two-mode free Fock space, $\Gamma^2 := \oplus_{n=0}^\infty \Gamma^2(n)$, where

$$\Gamma^2(n) := \mathbf{C}\text{-linear span}\{|\eta_1, \ldots, \eta_{2n}\rangle \mid \eta_i \in \{I, J\}\},$$

equipped with a canonical inner product

$$\langle \eta_1, \ldots, \eta_{2k} | \eta_1', \ldots, \eta_{2l}'\rangle = \delta_{(\eta_1, \ldots, \eta_{2k}), (\eta_1', \eta_{2l}')},$$

and operators a^+, a^- and a° specified by

$$a^+(|\eta_1, \ldots, \eta_{2n}\rangle) = |I, J, \eta_1, \ldots, \eta_{2n}\rangle + |J, I, \eta_1, \ldots, \eta_{2n}\rangle,$$

$$a^-|I, J, \eta_3, \ldots, \eta_{2n}\rangle = a^-|J, I, \eta_3, \ldots, \eta_{2n}\rangle = \frac{1}{2}|\eta_3, \ldots, \eta_{2n}\rangle,$$

$$a^-|I, I, \eta_3, \ldots, \eta_{2n}\rangle = a^-|J, J, \eta_3, \ldots, \eta_{2n}\rangle = 0,$$

$$a^\circ|I, J, \eta_3, \ldots, \eta_{2n}\rangle = \sqrt{\frac{\gamma}{2(1-\gamma)}} |J, J, \eta_3, \ldots, \eta_{2n}\rangle,$$

$$a^\circ|I, I, \eta_3, \ldots, \eta_{2n}\rangle = \sqrt{\frac{\gamma}{2(1-\gamma)}} |J, I, \eta_3, \ldots, \eta_{2n}\rangle,$$

$$a^\circ|J, I, \eta_3, \ldots, \eta_{2n}\rangle = \sqrt{\frac{1-\gamma}{2\gamma}} |I, I, \eta_3, \ldots, \eta_{2n}\rangle,$$

$$a^\circ|J, J, \eta_3, \ldots, \eta_{2n}\rangle = \sqrt{\frac{1-\gamma}{2\gamma}} |I, J, \eta_3, \ldots, \eta_{2n}\rangle. \quad (51)$$

Summing up the arguments above, we have established a central limit theorem associated with the decomposition (46) as follows.

Theorem 30 *Let $m \geq 1$ and $\epsilon_1, \ldots, \epsilon_m \in \{\pm, \circ\}$. Then for any constant $0 < \gamma < 1$, $u \in \Gamma^2$ and $\eta_1, \ldots, \eta_n \in \{I, J\}$, we have*

$$\lim_{N \to \infty} \langle u^{(N)}| \left(\frac{a_N^{\epsilon_1}}{\sqrt{v}}\right) \cdots \left(\frac{a_N^{\epsilon_m}}{\sqrt{v}}\right) |\eta_1, \ldots, \eta_n\rangle^{(N)}_{\Gamma(F_{2,\gamma}^{(N)})} = \langle u|a^{\epsilon_1} \cdots a^{\epsilon_m}|\eta_1, \ldots, \eta_n\rangle_{\Gamma^2},$$

where we use the notation (37). The operators a^+, a^- and a° are specified by (51). In particular, combinatorial moments with respect to the vacuum state *are given by*

$$\lim_{N\to\infty} \langle 0|\Big(\frac{1}{\sqrt{v}}\sum_{w_{ij}\in\Sigma_{2,\gamma}^{(N)}}\pi(w_{ij})\Big)^m|0\rangle_{\Gamma(F_{2,\gamma}^{(N)})}^{(N)} = \langle 0|(a^+ + a^- + a^\circ)^m|0\rangle_{\Gamma^2}. \quad (52)$$

Remarks.

(1) By a combinatorial argument, we see that the combinatorial moments (52) are given as the number of closed walks on a induced subgraph of a weighted binary tree.

(The weights $\lambda = 1/\sqrt{2}$ are given in the figure below.) Indeed m-th combinatorial moment F_m is given by the number of m-step walks which leave o and return to o. Let f_m be the number of m-step walks which leave z and return to z without arriving at o. By the self-similarity of the graph, we see for $m \geq 2$,

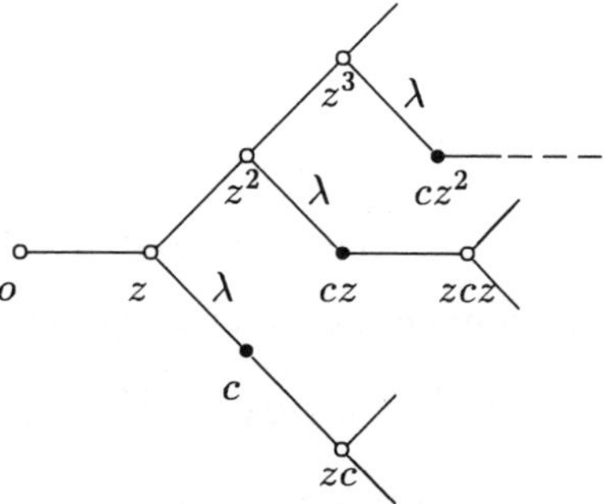

$$f_m = \sum_{k=0}^{m-2}\Big(f_k + \frac{F_k}{\sqrt{2}}\Big)f_{m-k-2}, \qquad F_m = \sum_{k=0}^{m-2} f_k F_{m-k-2},$$

where $f_0 = F_0 = 1$. Then the moment generating function $F(t) = \sum_m F_m t^m$ and $f(t) = \sum_m f_m t^m$ satisfy

$$f(t) - 1 = t^2\Big(f(t) + \frac{F(t)}{\sqrt{2}}\Big)f(t), \qquad F(t) - 1 = t^2 F(t)^2,$$

hence $t^2 F(t)^3 + t^2 F(t)^2 - 2F(t) + 2 = 0$. The Cauchy transform $G(t)$ of the distribution associated with the combinatorial moments (52) is given as a solution of $tG(t)^3 + G(t)^2 - 2tG(t) + 2 = 0$.

(2) The limit distribution associated with (52) coincides with the one in *Examples 1.5* (1.16) and (1.17) of Nica-Speicher[18], up to the variance, where the anti-commutation $ab + ba$ of semi-circular elements a, b which are free from each other is observed. Indeed we see that

$$\frac{1}{\sqrt{v}}\sum_{w_{ij}\in\Sigma_{2,\gamma}^{(N)}}\pi(w_{ij}) \approx \Big(\frac{\sigma_1 + \cdots + \sigma_{\gamma N}}{\sqrt{\gamma N}}\Big)\Big(\frac{\sigma_{\gamma N+1} + \cdots + \sigma_N}{\sqrt{(1-\gamma)N}}\Big)$$

$$+ \left(\frac{\sigma_{\gamma N+1} + \cdots + \sigma_N}{\sqrt{(1-\gamma)N}} \right) \left(\frac{\sigma_1 + \cdots + \sigma_{\gamma N}}{\sqrt{\gamma N}} \right).$$

which is nothing but the anti-commutation of semi-circular elements in the limit $N \to \infty$.

8.5 *Commutative association scheme and its quantum decomposition*

It is quite natural to apply our approach of the quantum decomposition to isotropic random walks on homogeneous graphs. In fact, it will be seen that any adjacency operator of a certain graph is decomposed into a sum of a 'creation', an 'annihilation' and a 'conservation' operators. Such a decomposition is motivated by the standard theory of stochastic evolutions[13]. This idea is useful for quantum limit theorems associated with random walks on graphs, particularly, on distance regular graphs. Indeed, Tabei[22] and Hashimoto-Obata-Tabei[11] carried out our idea on the Hamming scheme. A quantum decomposition of adjacency matrix of a Hamming graph is defined with the help of Euler's unicursal theorem, and a quantum central limit theorem is proved. As a result, we reproduce one of Hora's results[12] on the asymptotic distribution of eigen values of the adjacency matrix associated with a Hamming graph.

References

1. L. Accardi, M. Bożejko, *Interacting Fock spaces and Gaussianization of probability measures*, Infinite Dimen. Anal. Quantum Prob. **1** (1998), 663-670.
2. L. Accardi, Y. Hashimoto and N. Obata, *Notions of independence related to the free group*, Infinite Dimen. Anal. Quantum Prob. **1** (1998), 201-220.
3. L. Accardi, Y. Hashimoto and N. Obata, *Singleton independence*, Banach Center Publications **43** (1998), 9-24.
4. L. Accardi, Y. Hashimoto and N. Obata, *A role of singletons in quantum central limit theorems*, J. Korean Math. Soc. **35** (1998), 675-690.
5. L. Accardi and N. Obata, *Introduction to algebraic probability theory*, Nagoya Mathematical Lectures **2** (1999). (in Japanese)
6. K. Fichitner, W. Freudenberg and V. Liebscher, *Time evolution and invariance of boson systems beam splittings*, Infinite Dimen. Anal. Quantum Prob. **1** (1998), 511-531.

7. Y. Hashimoto, *A combinatorial approach to limit distributions of random walks: on discrete groups*, preprint, (1996).

8. Y. Hashimoto, *Deformations of the semi-circle law derived from random walks on free groups*, Prob. Math. Stat. **18** (1998), 399-410.

9. Y. Hashimoto, *Samples of algebraic central limit theorems based on* $\mathbf{Z}/2\mathbf{Z}$, In Infinite dimensional harmonic analysis, Transaction of a Japanese-German Symposium, Kyoto,1999, 115-126.

10. Y. Hashimoto, *Quantum decomposition in discrete groups and interacting Fock spaces*, Infinite Dimen. Anal. Quantum Prob. (2001) (to appear).

11. Y. Hashimoto, N. Obata and N. Tabei, *A quantum aspect of asymptotic spectral analysis of large Hamming graphs*, in "Quantum information IV (T. Hida and K. Saito Eds.)," (2001).

12. A. Hora, *Central limit theorems and asymptotic spectral analysis on large graphs*, Infinite Dimen. Anal. Quantum Prob. **1** (1998), 221-246.

13. R. Hudson and K. Parthasarathy, *Quantum Ito's formula and stochastic evolutions*, Commun. Math. Phys. **93** (1984),301-323.

14. P. Hilton and J. Pederson, *Catalan Numbers, Their Generalization and Their Uses*, Math. Intelligencer **13** (1991), 64-75.

15. Y. Lu: *On the interacting free Fock space and the deformed Wigner law*, Nagoya Math. J. **145** (1997), 1-28.

16. Y. Lu: *Interacting Fock spaces related to the Anderson model*, Infinite Dimensional Analysis and Quantum Probability **1** No.2 (1998), 247-283.

17. N. Muraki, *A new example of noncommutative "de Moivre-Laplace theorem"*, in "Probability Theory and Mathematical Statistics: Proceedings of the Seventh Japan-Russia Symposium (S. Watanabe, Ed.)," Tokyo, 1995, 353-362, World Scientific (1996).

18. A. Nica and R. Speicher, *Commutators of free random variables*, Duke Math. J. **92** No.3 (1998), 553-592.

19. F. Oravecz, *Deformed Poisson-laws as certain transform of deformed Gaussian-laws*, preprint (2000).

20. N. Saitoh and H. Yoshida, *Graphical representations of the q-creation and the q-annihilation operators and set partition statistics*, preprint (2000).

21. R. Speicher, *A new example of 'Independence' and 'White Noise'*, Prob. Th. Relat. Fields **84** (1990), 141-159.

22. N. Tabei, *Quantum decomposition on Hamming graphs and central limit theorem*, Master Thesis, Nagoya University (2001). (in Japanese)

23. D. Voiculescu, K. Dykema and A. Nica, *Free Random Variables*, CRM Monograph Ser. Amer. Math. Soc. (1992),

Quantum Information IV (pp. 79–86)
Eds. T. Hida and K. Saitô
© 2002 World Scientific Publishing Co.

SOME THOUGHTS ON THE INFINITE DIMENSIONAL HARMONIC ANALYSIS

TAKEYUKI HIDA

Faculty of Science and Technology, Meijo University
Nagoya 468-8502, Japan

YUKO HARA-MIMACHI

Department of Information Sciences, Meijo University
Nagoya 468-8502, Japan

White noise analysis has extensively developed during the past quarter century, and it is now able to think of the significant characteristics of the theory. Among others we should like to emphasize that the white noise analysis enjoys an aspect of an infinite dimensional harmonic analysis. In this article we shall discuss the role of the infinite dimensional rotation group in the study of white noise theory and propose some of future directions as well as its frontiers that shall be discussed in this line.

It is important for him who wants to discover not to confine himself to one chapter of science, but to keep in touch with others. - J. Hadamard -

Mathematics Subject Classification (2000): 60H40

1   Background

1.1   Rotation group

Let

$$E \subset L^2(R^d) \subset E^*$$

be a Gel'fand triple. The white noise measure μ is introduced on E^*. Let $O(E)$ be the group of linear isomorphisms g which are orthogonal. It is called the *infinite dimensional rotation group*. Let $O^*(E^*)$ be the collection of the adjoints g^* with $g \in O(E)$. The set $O^*(E^*)$ also forms a group which is isomorphic to $O(E)$ under the correspondence

$$g^{-1} \longrightarrow g^*.$$

There is an important property stated by the following

Proposition 1. The white noise measure μ is $O^*(E^*)$-invariant.

This assertion can be proved by the invariance of the characteristic functional of μ under the action of g, and the functional provides a link between the infinite dimensional rotation group and white noise analysis; indeed we can carry on the harmonic analysis arising from the group $O(E)$.

1.2 Fock space

We form a complex Hilbert space $(L^2) = L^2(E^*, \mu)$. A member of (L^2) is a *white noise functional*. The following assertion is known.

Proposition 2. The space (L^2) admits a direct sum decomposition:

$$(L^2) = \bigoplus_{n=0}^{\infty} H_n.$$

This is called the *Fock space*. A member of H_n is a homogeneous chaos of degree n.

1.3 Subgroups of $O(E)$

The group $O(E)$ is topologized by the compact-open topology. It is neither compact nor locally compact. In addition, as we shall see below, there are various subgroups of $O(E)$, each of which enjoys its own interesting property from the viewpoint of probability theory and/or analysis.

I. Finite dimensional subgroups of rotations.

Fix a complete orthonormal system $\{\xi_n\}$ of $L^2(R^d)$. Then, one can easily find a subgroup G_n of $O(E)$ such that G_n is isomorphic to $SO(n)$. Let G_∞ be the inductive limit of the G_n.

Associated with a member g of $O(E)$ is a unitary operator U_g acting on (L^2) which is defined by

$$(U_g\varphi)(x) = \varphi(g^*x).$$

The U_g reduces the subspace H_n. It is known that

Proposition 3.
1) The unitary representation $\{U_g, H_n\}$ is irreducible.
2) The subgroup G_∞ characterizes the infinite dimensional Laplace-Beltrami operator Δ_∞ which is given by

$$\Delta_\infty = \Sigma \{ \frac{\partial^2}{\partial \xi_n^2} - \langle x, \xi_n \rangle \frac{\partial}{\partial \xi_n} \}.$$

As a consequence the subspace H_n is the eigenspace of the infinite dimensional Laplace-Beltrami operator Δ_∞. In fact, the eigenvalue is $-n$.

II. The Lévy group $\mathcal{G}$.

Still a complete orthonormal system $\{\xi_n\}$ is fixed. Let π be a permutation of the natural numbers. Then, a change of the coordinate vectors is defined:

$$g_\pi : \quad \xi_n \longrightarrow \xi_{\pi(n)}.$$

Then, g_π extends to a linear transformation on E. Now assume that the density $d(\pi)$ of π is zero:

$$d(\pi) = \limsup \frac{1}{N} \#\{n \le N; \pi(n) > N\} = 0.$$

The collection

$$\mathcal{G} = \{g_\pi \in O(E); d(\pi) = 0\}$$

forms a group which is called the *Lévy group.*

The Lévy Laplacian defined by

$$\Delta_L = \lim \frac{1}{N} \sum_1^N \frac{\partial^2}{\partial \xi_n^2}$$

has a close connection with the Lévy group.

It is noted the following assertion.

Triviality The Lévy group has a continuous representation on the space of generalized white noise functionals.

III. Whiskers.

A *whisker* we mean that a one-parameter subgroup $\{g_t\}$ of $O(E)$ that comes from a one-parameter family of diffeomorphisms of the parameter space R^d. It is expressed in the form

$$g_t : \xi(u) \longrightarrow \xi(\psi_t(u)) \sqrt{|J(\psi_t(u))|},$$

where J means the Jacobian. There is assumed that

$$\psi_t \cdot \psi_s = \psi_{t+s}.$$

Example. The shift $\{S_t\}$. It is defined by

$$S_t \xi(u) = \xi(u - t).$$

Let T_t be the adjoint of S_t. Then $\{T_t, t \in R\}$ is the *flow of Brownian motion* after S. Kakutani [4]. His pioneering work has stimulated our study. Note that the shift describes the propagation of chaos as time t goes by.

The subgroups in II and III are, as it were, *essentially infinite dimensional*, although they have different flavour from each other. As for the subgroup III, we can see this property through the their spectrum; in general, they have countably Lebesgue spectrum.

2 Roles of whiskers

The plan of this report is to emphasize the important roles of whiskers, in particular conformal group, in the study of white noise analysis. In particular, special roles can be seen in the study of random fields which are indexed by a manifold and are formed by a white noise functional.

Let $\{g_t\}$ be a whisker as is given in the last section. Then, as is known, it is conjugate to the shift. More exactly, the function $\psi_t(u)$ defining the g_t can be expressed in the form

$$\psi_t(u) = f[f^{-1}(u) + t],$$

where f is a strictly monotone, continuous function on R. For important examples of a whisker with which we shall be concerned, the function f can be a function that is smooth enough. (See, e.g. [2, Chapt 5].)

We now give an interpretation why a whisker is essentially infinite dimensional. Remind a notion of the average power $r(x, g)$ of a rotation g. The definition is (see [1])

$$r(x,g) = \limsup \frac{1}{n} \sum_{k=1}^{n} \langle x, \xi_k - g\xi_k \rangle^2.$$

Note that this definition depends on the choice of a complete orthonormal system.

If the average power is positive, then g is considered to be *essentially infinite dimensional*.

Proposition 4. For the shift $\{S_t\}$, any S_t, with $t \neq 0$ is essentially infinite dimensional.

Proof. Take an S_t. Choose a complete orthonormal system $\{\xi_k\}$. We have an equality

$$\langle x, S_t\xi\rangle = \langle \hat{x}, \exp[i\lambda t]\hat{\xi}\rangle,$$

from which we can easily prove the assertion.

Theorem 1. All the nontrivial whiskers are essentially infinite dimensional.

Proof. From Proposition 4 and the fact that any whisker is conjugate to the shift except trivial cases, we can prove the theorem.

Remark. There are also many other members in the Lévy group which are essentially infinite dimensional.

Coming back to whisker, it is convenient to have its infinitesimal generator α, which can be expressed in the form

$$\alpha = a(u)\frac{d}{du} + \frac{1}{2}a'(u),$$

with a suitable function $a(u)$.

With the help of the generators we can discover complex relationship between whiskers, and even think of their roles in white noise analysis.

3 Conformal group

Starting from the shift, which describes the flow of Brownian motion (and hence the most important whisker), we can find some other whiskers which have good commutation relations and have significant probabilistic meaning.

Actually, we are given several interesting whiskers and putting them together it is proved that they form a conformal group, denoted by $C(d)$. The parameter space of white noise is now taken to be R^d.

With some generalization of what has been discussed in [2], we claim that those whiskers in question can be listed as follows:

1) Shift,

2) isotropic dilations,

3) subgroup isomorphic to $SO(d)$,

4) special conformal transformations.

It is known that the subgroup $C(d)$ of $O(E)$ is isomorphic to the classical group $SO(d+1,1)$. Its Lee group structure help our study of white noise.

One of the interesting and in fact significant applications is that a unitary representation of the group $C(d)$ contributes to the innovation theory of a random field.

Let $X(C)$ be a random field parameterized by a closed convex manifold C. Assume, in particular, that the parameter C runs through a class $\mathbf{C}$ of ovaloids in R^d. We are interested in the case where $X(C)$ is expressed as a causal (generalized) functional of white noise $x(u), u \in (C)$, where (C) denites a domain enclosed by C. Generally speaking, geometric structure of the manifold C can describe the information theoretical behavior of the random field $X(C)$. The C, unlike the time parameter t, enjoys complex structure, so that when C deforms $X(C)$ carries more information than $X(t)$. In particular, the variation $\delta X(C)$ by using conformal transformations is useful for our study. The basic idea is illustrated below.

The infinitesimal deformation of C is represented by a family of functions $\{\delta(s), s \in C\}$. We apply the conformal transformations that leave the C invariant. Then, we are given various family of functions $\delta(s)$ which is dense in the L^2-class of functions over C. This comes from the irreducible unitary represrentation theory of the conformal group. This claims that the $\{x(s), s \in C\}$ can be obtained if $X(C)$ is Gaussian. The system is the innovation for $X(C)$.

We assume that C is taken as before and and the representation of $X(C)$ is causal in $x(u)$. Then, we can prove

Theorem 2. For a Gaussian random fields $X(C)$, the innovation is obtained by the conformal transformations acting on the basic parameter space R^d.

Further we can introduce the so-called stochastic variational equation which determines a random field. See [8].

4 Concluding remarks

There are three remarks.

I. Reversibility.

As soon as we come to the study of a random field $X(C)$, we have to worry about how to define the reversibility of the random phenomena. For $X(t)$ only one direction and reverse can be considered when we think of propagation of the random phenomena. As for the $X(C)$ the direction of propagation is not unique, but should not vaguely defined. Deformations of the parameter C are recommended to be taken within the conformal group.

Here is a useful information. In the two dimensional parameter case, we prefer to use notations in the complex plane. A conformal mapping from unit circle onto itself is written as

$$w = \gamma \frac{z - \alpha}{\bar{\alpha} z - 1},$$

where $|\gamma| = 1$.

II. Applications to physics.

Stochastic variational equation determines a Euclidean field. We hope this can be extended to relativistic quantum fields. We have hope to establish a path integral for a random field indexed by C. This would be a good approach to quantization of a quantum field.

III. The white noise theory has so far much related (not based on) to the so-called L^2-nonlinearity. It should be refeminded that the original idea of the white noise theory came from, as it were, the *reductionism*; nemely, we start with a system of basic and atomic random variables, like white noise, and analyse the functionals of those variable in the system. The assumption to restrict functionals in L^2-class is somewhat too restrictive. There are many other interesting classes of random functions which have not variances, so that the analysis requires suitable topology which is not the mean square, and hence it needs new tools from analysis, like entropy. Anyhow, we should now think of some new frontier of white noise analysis. The authors have already started a study in this direction and given lectures at the present Academic Frontier Project Seminar. Some part of our study will also be printed as the literature [3]. Further development will be reported in the Quantum Information Series.

References

[1] L. Accardi et al ed. Selected papers of Takeyuki Hida. World Scientific Pub. Co. 2001.

[2] T. Hida, Brownian motion. 1975 (in Japanese); English Translation 1980 Springer-Verlag.

[3] T. Hida, White noise analysis: A new frontier, Univ. Roma, Volterra Center Pub. N. 499, 2002.

[4] S. Kakutani, Determination of the spectrum of the flow of Brownian motion. Proc. National Acad. Sci., 36 (1950), 319-323.

[5] H.-H. Kuo, White noise distribution theory, 1996, CRC Press.

[6] P. Lévy, Problèmes concrets d'analyse fonctionnelle. 1951 Gauthier-Villars.

[7] P. Lévy, Processus stochastiques et mouvement brownien. 1948: 2ème ed. 1965, Gauthier-Villars.

[8] Si Si, Innovation for some random fields. J. Korean Math. Soc. 35, no.3, 1998, 793-802.

[9] Si Si, Gaussian processes and Gaussian random fields. Quantum Information II2000, World Scientific Pub. Co., pp 195-204.

[10] Si Si and Win Win Htay, Topics on Complex Gaussian random fields, 2001. Quantum Information IV, World Scientific Pub. Co., (in this volume).

Quantum Information IV (pp. 87–102)
Eds. T. Hida and K. Saitô
© 2002 World Scientific Publishing Co.

A TREATMENT OF QUANTUM BAKER'S MAP BY CHAOS DEGREE

KEI INOUE, MASANORI OHYA

Department of Information Sciences
Science University of Tokyo
Noda City, Chiba 278-8510 Japan

IGOR V. VOLOVICH

Steklov Mathematical Institute,
Russian Academy of Science
Gubkin St. 42, Moscow, GSP1, 117966,
Russia

We study the chaotic behaviour and the quantum-classical correspondence for the baker's map. Correspondence between quantum and classical expectation values is investigated and it is shown that it is lost at the logarithmic timescale. The quantum chaos degree is computed and it is demonstrated that it describes the chaotic features of the model. The correspondence between classical and quantum chaos degrees is considered.

1 Introduction

The study of chaotic behaviour in classical dynamical systems is dating back to Lobachevsky and Hadamard who have been studied the exponential instability property of geodesics on manifolds of negative curvature and to Poincare, who initiated the inquiry into the stability of the solar system. One believes now that the main features of chaotic behaviour in the classical dynamical systems are rather well understood, see for example [1, 27]. However the status of "quantum chaos" is much less clear although the significant progress has been made on this front.

Sometimes one says that an approach to quantum chaos, which attempts to generalize the classical notion of sensitivity to initial conditions, fails for two reasons: first there is no quantum analogue of the classical phase space trajectories and, second, the unitarity of linear Schrodinger equation precludes sensitivity to initial conditions in the quantum dynamics of state vector. Let us remind, however, that in fact there exists a quantum analogue of the classical phase space trajectories. It is quantum evolution of expectation values of appropriate observables in suitable states. Also let us remind that the dynamics of a classical system can be described either by the Hamilton equations or by the liner Liouville equations. In quantum theory the linear

Schrodinger equation is the counterpart of the Liouville equation while the quantum counterpart of the classical Hamilton's equation is the Heisenberg equation. Therefore the study of quantum expectation values should reveal the chaotic behaviour of quantum systems. In this paper we demonstrate this fact for the quantum baker's map.

If one has the classical Hamilton's equations

$$dq/dt = p, \quad dp/dt = -V'(q),$$

then the corresponding quantum Heisenberg equations have the same form

$$dq_h/dt = p_h, \quad dp_h/dt = -V'(q_h),$$

where q_h and p_h are quantum canonical operators of position and momentum. For the expectation values one gets the Ehrenfest equations

$$d < q_h > /dt = < p_h >, \quad d < p_h > /dt = - < V'(q_h) >$$

Note that the Ehrenfest equations are classical equations but for non non-linear $V'(q_h)$ they are neither Hamiltonian's equations nor even differential equations because one can not write $< V'(q_h) >$ as a function of $< q_h >$ and $< p_h >$. However these equations are very convenient for the consideration of the semiclassical properties of quantum system. The expectation values $< q_h >$ and $< p_h >$ are functions of time and initial data. They also depend on the quantum states. One of important problems is to study the dependence of expectation values from the initial data. In this paper we will study this problem for the quantum baker's map.

The main objective of "quantum chaos" is to study the correspondence between classical chaotic systems and their quantum counterparts in the semi-classical limit [8,7]. The quantum-classical correspondence for dynamical systems has been studied for many years, see for example [3,10,30] and reference therein. A significant progress in understanding of this correspondence has been achieved in the WKB approach when one considers the Planck constant h as a small variable parameter. Then it is well known that in the limit $h \to 0$ quantum theory is reduced to the classical one. However in physics the Planck constant is a fixed constant although it is very small. Therefore it is important to study the relation between classical and quantum evolutions when the Planck constant is fixed. There is a conjecture [29,5] that a characteristic timescale τ appears in the quantal evolution of chaotic dynamical systems. For time less then τ there is a correspondence between quantum and classical expectation values, while for times greater that τ the predictions of the classical and quantum dynamics no longer coincide. The important problem is to estimate the dependence τ on the Planck constant h. Probably a universal formula expressing τ in terms of h does not exist and every model should be studied case by case. It is expected that certain quantum and classical expectation values diverge on a timescale inversely proportional to some power of h [4]. Other authors suggest that a breakdown may be anticipated on a much smaller logarithmic timescale [20,9,22,14,17,24,25,18]. The characteristic time τ associated with the hyperbolic fixed points of the classical motion is expected to be of the logarithmic form $\tau = \frac{1}{\lambda} \ln \frac{C}{h}$ where λ is the Lyapunov exponent and C is a constant which can be taken to be the classical action. Such the logarithmic timescale has been found in the numerical simulations of some dynamical models [30]. It was shown also that the discrepancy between quantum and classical evolutions is decreased by even a small coupling with the environment, which in the quantum case leads to decoherence [30].

The chaotic behaviour of the classical dynamical systems is often investigated by computing the Lyapunov exponents. An alternative quantity measuring chaos in dynamical systems which is called the chaos degree has been suggested in [19] in the general framework of information dynamics [12]. The chaos degree was applied to various models in [13]. An advantage of the chaos degree is that it can be applied not only to classical systems but also to quantum systems as well.

In this work we study the chaotic behaviour and the quantum-classical correspondence for the baker's map [4,21]. The quantum baker's map is a simple model invented for the theoretical study of quantum chaos. Its mathematical properties have been studied in numerical works. In particular its semiclassical properties have been considered [20,9,22,14,17,24,25,18], quantum computing and optical realizations have been proposed [11,23,6], various quantization procedures have been discussed [16,15,22,26], a symbolic dynamics representation has been given [26].

It is well known that for the consideration of the semiclassical limit in quantum mechanics it is very useful to use coherent states. We define an analogue of the coherent states for the quantum baker's map. We study the quantum baker's map by using the correlation functions of the special form which corresponds to the expectation values of Weyl operators, translated in time by the unitary evolution operator and taken in the coherent states. To explain our formalism we first discuss the classical limit for correlation functions in ordinary quantum mechanics. Correspondence between quantum and classical expectation values for the baker's map is investigated and it is shown that it is lost at the logarithmic timescale. The chaos degree for the quantum baker's map is computed and it is demonstrated that it describes the chaotic features of the model. The dependence of the chaos degree on the Planck constant is studied and the correspondence between classical and quantum chaos degrees is established..

2 Quantum vs. Classical Dynamics

In this section we discuss an approach to the semiclassical limit in quantum mechanics by using the coherent states, see [10]. Then in the next section an extension of this approach to the quantum baker's map will be given. Consider the canonical system with the Hamiltonian function

$$H = \frac{p^2}{2} + V(x) \tag{1}$$

in the plane $(p, x) \in \mathbf{R}^2$. We assume that the canonical equations

$$\dot{x}(t) = p(t), \qquad \dot{p}(t) = -V'(x(t)) \tag{2}$$

have a unique solution $(x(t), p(t))$ for times $|t| < T$ with the initial data

$$x(0) = x_0, \qquad p(0) = v_0 \tag{3}$$

This is equivalent to the solution of the Newton equation

$$\ddot{x}(t) = -V'(x(t)) \tag{4}$$

with the initial data

$$x(0) = x_0, \qquad \dot{x}(0) = v_0 \tag{5}$$

We denote

$$\alpha = \frac{1}{\sqrt{2}}(x_0 + iv_0) \tag{6}$$

The quantum Hamiltonian operator has the form

$$H_h = \frac{p_h^2}{2} + V(q_h)$$

where p_h and q_h satisfy the commutation relations

$$[p_h, q_h] = -ih$$

The Heisenberg evolution of the canonical variables is defined as

$$p_h\left(t\right) = U\left(t\right)p_h U\left(t\right)^*, \qquad q_h\left(t\right) = U\left(t\right)q_h U\left(t\right)^*$$

where

$$U\left(t\right) = \exp\left(-itH_h/h\right)$$

For the consideration of the classical limit we take the following representation

$$p_h = -ih^{1/2}\partial/\partial x, \qquad q_h = h^{1/2}x$$

acting to functions of the variable $x \in \mathbf{R}$. We also set

$$a = \frac{1}{\sqrt{2}h^{1/2}}\left(q_h + ip_h\right) = \frac{1}{\sqrt{2}}\left(x + \frac{\partial}{\partial x}\right),$$

$$a^* = \frac{1}{\sqrt{2}h^{1/2}}\left(q_h - ip_h\right) = \frac{1}{\sqrt{2}}\left(x - \frac{\partial}{\partial x}\right),$$

then

$$[a, a^*] = 1.$$

The coherent state $|\alpha\rangle$ is defined as

$$|\alpha\rangle = W\left(\alpha\right)|0\rangle \tag{7}$$

where α is a complex number, $W\left(\alpha\right) = \exp\left(\alpha a^* - a\alpha^*\right)$ and $|0\rangle$ is the vacuum vector, $a\,|0\rangle = 0$. The vacuum vector is the solution of the equation

$$\left(q_h + ip_h\right)|0\rangle = 0 \tag{8}$$

In the x - representation one has

$$|0\rangle = \exp\left(-x^2/2\right)/\sqrt{2\pi}. \tag{9}$$

The operator $W\left(\alpha\right)$ one can write also in the form

$$W\left(\alpha\right) = Ce^{iq_h v_0/h^{1/2}}e^{-ip_h x_0/h^{1/2}} \tag{10}$$

where $C = \exp\left(-v_0 x_0/2h\right).$

The mean value of the position operator with respect to the coherent vectors is the real valued function

$$q\left(t,\alpha,h\right) = \left\langle h^{-1/2}\alpha \left| q_h\left(t\right) \right| h^{-1/2}\alpha \right\rangle \tag{11}$$

Now one can present the following basic formula describing the semiclassical limit

$$\lim_{h\to 0} q\left(t,\alpha,h\right) = x\left(t,\alpha\right) \tag{12}$$

Here $x\left(t,\alpha\right)$ is the solution of (4) with the initial data (5) and α is given by (6).

Let us notice that for time $t = 0$ the quantum expectation value $q\left(t,\alpha,h\right)$ is equal to the classical one:

$$q\left(0,\alpha,h\right) = x\left(0,\alpha\right) = x_0 \tag{13}$$

for any h. We are going to compare the time dependence of two real functions $q\left(t,\alpha,h\right)$ and $x\left(t,\alpha\right)$. For small t these functions are approximately equal. The important problem is to estimate for which t the large difference between them will appear. It is expected that certain quantum and classical expectation values diverge on a timescale inversely proportional to some power of h [10]. Other authors suggest that a breakdown may be anticipated on a much smaller logarithmic timescale [20,9,22,14,17,24,25,18]. One of very interesting examples of classical systems with chaotic behaviour is described by the hamiltonian function

$$H = \frac{p_1^2}{2} + \frac{p_2^2}{2} + \lambda x_1^2 x_2^2$$

The consideration of this classical and quantum model within the described framework will be presented in another publication.

3   Coherent States for the Quantum Baker's Map

The classical baker's transformation maps the unit square $0 \le q,p \le 1$ onto itself
according to

$$(q,p) \to \begin{cases} (2q,p/2), & \text{if } 0 \le q \le 1/2 \\ (2q-1,(p+1)/2), & \text{if } 1/2 < q \le 1 \end{cases}$$

This corresponds to compressing the unit square in the p direction and stretching it in the q direction, while preserving the area, then cutting it vertically and stacking the right part on top of the left part.

To quantize the unite square one defines the unitary displacement operators $\hat{U}$ and $\hat{V}$ in D - dimensional Hilbert space, which produce displacements in the momentum and position directions, respectively, and which obey the commutation relation

$$\hat{U}\hat{V} = \epsilon\hat{V}\hat{U},$$

where $\epsilon = \exp(2\pi i/D)$. We choose $D = 2^N$. The "Planck constant" $h = 1/D = 2^{-N}$. The operators $\hat{U}$ and $\hat{V}$ can be written as

$$\hat{U} = e^{2\pi i\hat{q}}, \quad \hat{V} = e^{2\pi i\hat{p}}$$

The position and momentum operators $\hat{q}$ and $\hat{p}$ both have eigenvalues j/D, $j = 0, ..., D-1$. If $\{|q_j\rangle\}$ are the eigenvectors of the position operator $\hat{q}$ then the eigenvectors of the momentum operator $\{|p_j\rangle\}$ are obtained by using the discrete Fourier transform F_N :

$$|p_j\rangle = F_N |q_j\rangle = \frac{1}{\sqrt{D}} \sum_{k=0}^{D-1} e^{2\pi ikj/D} |q_j\rangle$$

The quantum baker's map is written as the following matrix

$$T = F_N^{-1} \begin{pmatrix} F_{N-1} & \\ & F_{N-1} \end{pmatrix} \tag{14}$$

We define the coherent states by

$$|\alpha\rangle = Ce^{2\pi i\hat{q}v}e^{-2\pi i\hat{p}x} |\psi_0\rangle \tag{15}$$

Here $\alpha = x + iv$, x and v are integers, C is the normalization constant and $|\psi_0\rangle$ is the vacuum vector. This definition should be compared with 10. The vacuum vector can be defined as the solution of the equation

$$(q_h + ip_h) |\psi_0\rangle = 0$$

(compare with 8). We will use the simpler definition which in the position representation is

$$\langle q_j |\psi_0\rangle = C \exp\left(-q_j^2/2\right)$$

(compare with 9). Here C is a normalization constant.

4 Chaos Degree

Let us review the entropic chaos degree defined in [19]. This entropic chaos degree is given by the probability distribution φ of a orbit generated by a dynamics (channel) Λ^* sending a state to a state; $\varphi = \sum_k p_k \delta_k$, where δ_k is the delta measure such as $\delta_k(j) \equiv \begin{cases} 1 \ (k=j) \\ 0 \ (k \neq j) \end{cases}$. Then the entropic chaos degree is defined as

$$D\left(\varphi; \Lambda^*\right) = \sum_k p_k S(\Lambda^* \delta_k) \tag{16}$$

with the von Neumann entropy S, equivalently to the Shannon entropy because the probability distribution φ is a classical object.

The above channel Λ^* produces a dynamics F of the orbit, so that let x_n be the orbit or a certain function of the orbit at time n and F be a map from x_n to x_{n+1}, For a map F on $I \equiv [a,b]^N \subset \mathbf{R}^N$ with $x_{n+1} = F(x_n)$ (a difference equation), let $I \equiv \bigcup_k B_k$ be a finite partition with $B_i \cap B_j = \emptyset \ (i \neq j)$. The state $\varphi^{(n)}$ of the orbit determined by the difference equation is defined by the probability distribution $\left(p_i^{(n)}\right)$, that is, $\varphi^{(n)} = p^{(n)} = \sum_i p_i^{(n)} \delta_i$, where for an initial value $x \in I$ and the characteristic function 1_A

$$p_i^{(n)} \equiv \frac{1}{m+1} \sum_{k=n}^{m+n} 1_{B_i}\left(F^k x\right).$$

When the initial value x is distributed due to a measure v on I, the above $p_i^{(n)}$ is given as

$$p_i^{(n)} \equiv \frac{1}{m+1} \int_I \sum_{k=n}^{m+n} 1_{B_i}\left(F^k x\right) d\nu.$$

The joint distribution $\left(p_{ij}^{(n,n+1)}\right)$ between the time n and $n+1$ is defined by

$$p_{ij}^{(n,n+1)} \equiv \frac{1}{m+1} \sum_{k=n}^{m+n} 1_{B_i}\left(F^k x\right) 1_{B_j}\left(F^{k+1} x\right)$$

or

$$p_{ij}^{(n,n+1)} \equiv \frac{1}{m+1} \int_I \sum_{k=n}^{m+n} 1_{B_i}\left(F^k x\right) 1_{B_j}\left(F^{k+1} x\right) d\nu.$$

Then the channel Λ_n^* at n is determined by

$$\Lambda_n^* \equiv \left(\frac{p_{ij}^{(n,n+1)}}{p_i^{(n)}}\right) \implies p^{(n+1)} = \Lambda_n^* p^{(n)},$$

and the chaos degree is given by

$$D_c\left(p^{(n)}; \Lambda_n^*\right) = \sum_i p_i^{(n)} S(\Lambda_n^* \delta_i) = \sum_{i,j} p_{ij}^{(n,n+1)} \log \frac{p_i^{(n)}}{p_{ij}^{(n,n+1)}}. \qquad (17)$$

This classical chaos degree was applied to several dynamical maps such logistic map, Baker's transformation and Tinkerbel map, and it could explain their chaotic characters [19,13]. Our chaos degree has several merits compared with usual measures such as Lyapunov exponent.

We can judge whether the dynamics causes a chaos or not by the value of D as

$$D > 0 \iff \text{chaotic},$$
$$D = 0 \iff \text{stable}.$$

5 Chaos Degree for the Quantum Baker's map

In this section, we show a general representation of the mean value of the position operator $\hat{q}$ for the time evolution, which is constructed by the quantum baker's map. Then we give the algorithm to compute the chaos degree for the quantum baker's map.

Recently a whole class of quantum baker's maps has been defined in [26]. It is a quantum analogy of the symbolic dynamics [2] for the classical baker's map.

For any n, $1 \leq n \leq N - 2$, a quantum baker's map T_n can be defined by

$$T_n \left|\xi_1\xi_2 \cdots \xi_n \bullet \xi_{n+1}\xi_{n+2} \cdots \xi_N\right\rangle \equiv \left|\xi_1\xi_2 \cdots \xi_{n+1} \bullet \xi_{n+2}\xi_{n+3} \cdots \xi_N\right\rangle, \qquad (18)$$

where the vector $\left|\xi_1\xi_2 \cdots \xi_n \bullet \xi_{n+1}\xi_{n+2} \cdots \xi_N\right\rangle$ is given by

$$\left|\xi_1\xi_2 \cdots \xi_n \bullet \xi_{n+1}\xi_{n+2} \cdots \xi_N\right\rangle \equiv \left|\xi_{n+1}\right\rangle \otimes \cdots \otimes \left|\xi_N\right\rangle e^{i\pi(0.\xi_n \cdots \xi_1 1)} \otimes$$
$$\sqrt{1/2}\left\{|0\rangle + \exp\left[2\pi i\left(0.\xi_1 1\right)\right]|1\rangle\right\} \otimes$$
$$\sqrt{1/2}\left\{|0\rangle + \exp\left[2\pi i\left(0.\xi_2\xi_1 1\right)\right]|1\rangle\right\} \otimes \cdots \otimes$$
$$\sqrt{1/2}\left\{|0\rangle + \exp\left[2\pi i\left(0.\xi_n \cdots \xi_1 1\right)\right]|1\rangle\right\}$$

for n, $1 \leq n \leq N-1$ [26]. For $n = N-1$, the map is the original baker's map (14) as defined in [4] and it becomes

$$T_{N-1} \left| \xi_1 \xi_2 \cdots \xi_{N-1} . \xi_N \right\rangle \equiv \left| \xi_1 \xi_2 \cdots \xi_N \right\rangle . \tag{19}$$

For $n = 0$, the map is

$$T_0 \left| . \xi_1 \xi_2 \cdots \xi_N \right\rangle \equiv \left| \xi_1 . \xi_2 \xi_3 \cdots \xi_N \right\rangle . \tag{20}$$

To study the time evolution and the classical limit $h \to 0$ which corresponds to $N \to \infty$ of the quantum baker's map T_0, we introduce the following the mean value of the position operator $\hat{q}$ for time $n \in \mathbf{N}$ with respect to a single basis$|\xi\rangle$:

$$r_N^{(n)} = \langle \xi | \, T_0^n \hat{q} T_0^{-n} \, | \xi \rangle , \tag{21}$$

where $|\xi\rangle = |\xi_1 \xi_2 \cdots \xi_N\rangle$ and $\hat{q} = \sum_{j=0}^{2^N - 1} q_j |j\rangle \langle j|$ with eigen values $q_j = \frac{j+1/2}{2^N}$, $j = 0, 1, \ldots, 2^N - 1$, $j = \sum_{k=1}^{N} j_k 2^{N-k}$ and $j_k = 0, 1$.

The simple formula of the matrix elements of T_0 with respect to two different bases is given by

$$\langle \xi^0 | \, T_0 \, | \xi^1 \rangle = \frac{1-i}{2} \prod_{k=2}^{N} \delta \left(\xi_k^0 - \xi_{k-1}^1 \right) \exp \left(\frac{\pi}{2} i \left| \xi_1^0 - \xi_N^1 \right| \right) , \tag{22}$$

where$|\xi^0\rangle = |\xi_1^0 \xi_2^0 \cdots \xi_N^0\rangle$ and $|\xi^1\rangle = |\xi_1^1 \xi_2^1 \cdots \xi_N^1\rangle$ [28]. From (22), the following formula of the matrix elements of T_0^n for any $n \in \mathbf{N}$ is easily obtained.

$$\langle \xi^0 | \, T_0^n \, | \xi^1 \rangle$$
$$= \begin{cases} \left(\frac{1-i}{2} \right)^n \left(\prod_{k=1}^{N-n} \delta \left(\xi_{n+k}^0 - \xi_k^1 \right) \right) \left(\prod_{l=1}^{n} A_{\xi_l^0 \xi_{N-n+l}^1} \right) & \text{if } n < N \\ \left(\frac{1-i}{2} \right)^n \left(\prod_{k=1}^{n} A_{\xi_k^0 \xi_k^1} \right) & \text{if } n = N \\ \left(\frac{1-i}{2} \right)^n \left(\prod_{k=1}^{p} \left(A^{m+1} \right)_{\xi_k^0 \xi_{N-p+k}^1} \right) \left(\prod_{l=1}^{N-p} \left(A^m \right)_{\xi_{p+l}^0 \xi_l^1} \right) & \text{if } n = mN + p \\ \left(\frac{1-i}{2} \right)^n \prod_{k=1}^{N} \left(A^m \right)_{\xi_k^0 \xi_k^1} & \text{if } n = mN, \end{cases} \tag{23}$$

where$|\xi^0\rangle = |\xi_1^0 \xi_2^0 \cdots \xi_N^0\rangle$, $|\xi^1\rangle = |\xi_1^1 \xi_2^1 \cdots \xi_N^1\rangle$, A is the 2×2 matrix with the element $A_{x_1 x_2} = \exp \left(\frac{\pi}{2} i \left| x_1 - x_2 \right| \right)$ for $x_1, x_2 = 0$,1, $p = 1, \cdots, N-1$ and $m \in \mathbf{N}$.

Using these formula, the following theorem are proved.

THEOREM 5.1 $r_N^{(n)} =$

$$
\begin{cases}
\sum_{k=1}^{N-n} \xi_{n+k} 2^{-k} + \frac{2^n}{2^{N+1}} & \text{if } n < N \\
\frac{1}{2} & \text{if } n = N \\
\frac{1}{2^n} \sum_{j=0}^{2^N-1} \frac{j+1/2}{2^N} \prod_{k=1}^{p} \left| (A^{m+1})_{\xi_k \ j_{N-p+k}} \right|^2 \prod_{l=1}^{N-p} \left| (A^m)_{\xi_{p+l} \ j_l} \right|^2 & \text{if } n = mN + p \\
\frac{1}{2^n} \sum_{j=0}^{2^N-1} \frac{j+1/2}{2^N} \prod_{k=1}^{N} \left| (A^m)_{\xi_k \ j_k} \right|^2 & \text{if } n = mN,
\end{cases}
\tag{24}
$$

where $\left| \xi^0 \right\rangle = \left| \xi_1^0 \xi_2^0 \cdots \xi_N^0 \right\rangle$, $\left| \xi^1 \right\rangle = \left| \xi_1^1 \xi_2^1 \cdots \xi_N^1 \right\rangle$, A is the 2×2 matrix with the element $A_{x_1 x_2} = \exp\left(\frac{\pi}{2} i \left| x_1 - x_2 \right| \right)$ for $x_1, x_2 = 0$,1, $p = 1, \cdots, N-1$ and $m \in \mathbf{N}$.

By diagonalizing the matrix A, we obtain the following formula of the absolute square of the matrix elements of A^n for any $n \in \mathbf{N}$.

LEMMA 5.2 For any $n \in \mathbf{N}$, we have

$$
\left| (A^n)_{kj} \right|^2 = \begin{cases}
2^n \cos^2 \left(\frac{n\pi}{4} \right) & \text{if } k = j \\
2^n \sin^2 \left(\frac{n\pi}{4} \right) & \text{if } k \neq j
\end{cases}.
$$

Combining the above theorem and lemma, we obtain the following two theorems with respect to the mean value $r_N^{(n)}$ of the position operator.

THEOREM 5.3 For the case $n = mN + p$, $p = 1, 2, \ldots N - 1$ and $m \in \mathbf{N}$, we have

$$
r_N^{(n)} = \begin{cases}
\sum_{k=1}^{N-p} \xi_{p+k} 2^{-k} + \frac{2^p}{2^{N+1}} & \text{if } m = 0 \ (mod \ 4) \\
\sum_{k=N-p+1}^{N} \eta_{k-(N-p)} 2^{-k} + \frac{2^N - 2^p + 1}{2^{N+1}} & \text{if } m = 1 \ (mod \ 4) \\
\sum_{k=1}^{N-p} \eta_{p+k} 2^{-k} + \frac{2^p}{2^{N+1}} & \text{if } m = 2 \ (mod \ 4) \\
\sum_{k=N-p+1}^{N} \xi_{k-(N-p)} 2^{-k} + \frac{2^N - 2^p + 1}{2^{N+1}} & \text{if } m = 3 \ (mod \ 4),
\end{cases}
\tag{25}
$$

where $\eta_k = \xi_k + 1 \ (mod \ 2)$, $k = 1, \cdots, N$.

THEOREM 5.4 For the case $N = mN, m \in \mathbf{N}$, we have

$$
r_N^{(n)} = \begin{cases}
\sum_{k=1}^{N} \xi_k 2^{-k} + \frac{1}{2^{N+1}} & \text{if } m = 0 \ (mod \ 4) \\
\frac{1}{2} & \text{if } m = 1, 3 \ (mod \ 4) \\
\sum_{k=1}^{N} \eta_k 2^{-k} + \frac{1}{2^{N+1}} & \text{if } m = 2 \ (mod \ 4).
\end{cases}
\tag{26}
$$

Using these formulas (24), (25) and (26), the probability distribution $\left(p_i^{(n)} \right)$ of the orbit of mean value $r_N^{(n)}$ of the position operator $\hat{q}$ for the time evolution, which is constructed by the quantum baker's map, is given by

$$p_i^{(n)} \equiv \frac{1}{m+1} \sum_{k=n}^{m+n} 1_{B_i}\left(r_N^{(k)}\right)$$

for an initial value $r_N^{(0)} \in [0,1]$ and the characteristic function 1_A. The joint distribution $\left(p_{ij}^{(n,n+1)}\right)$ between the time n and $n+1$ is given by

$$p_{ij}^{(n,n+1)} \equiv \frac{1}{m+1} \sum_{k=n}^{m+n} 1_{B_i}\left(r_N^{(k)}\right) 1_{B_j}\left(r_N^{(k+1)}\right).$$

Thus the chaos degree for the quantum baker's map is calculated by

$$D_q\left(p^{(n)};\Lambda_n^*\right) = \sum_{i,j} p_{ij}^{(n,n+1)} \log \frac{p_i^{(n)}}{p_{ij}^{(n,n+1)}}, \tag{27}$$

whose numerical value is shown in the next section.

6 Numerical Simulation of the Chaos Degree and Classical-Quantum Correspondence

We compare the dynamics of the mean value $r_N^{(n)}$ of position operator $\hat{q}$ with that of the classical value $q^{(n)}$ in the q direction. We take an initial value of the mean value as

$$r_N^{(0)} = \sum_{l=1}^{N} \xi_l 2^{-l} + 1/2^{N+1} = 0.\xi_1\xi_2\cdots\xi_N 1,$$

where ξ_i is a pseudo-random number valued with 0 or 1. At the time zero we assume that the classical value $q^{(0)}$ in the q direction takes the same value as the mean value $r_N^{(0)}$ of position operator $\hat{q}$. The distribution of $r_N^{(n)}$ for the case $N = 500$ is shown in Fig.1 up to the time $n = 1000$. The distribution of the classical value $q^{(n)}$ for the case $N = 500$ in the q direction is shown in Fig.2 up to the time $n = 1000$.

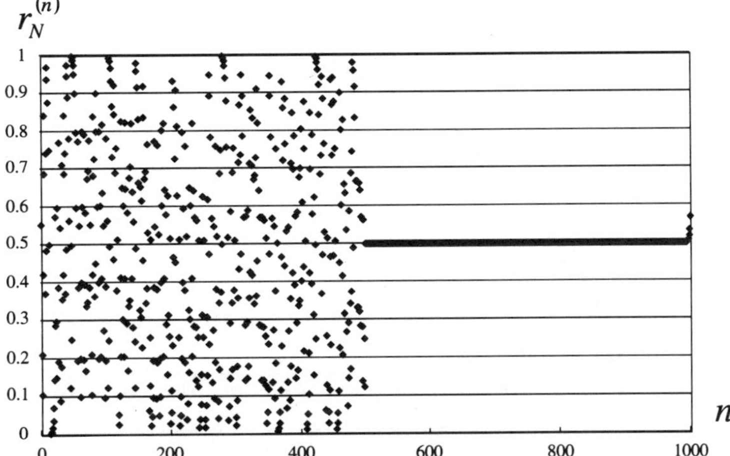

Fig.1. The distribution of $r_N^{(n)}$ for the case N=500

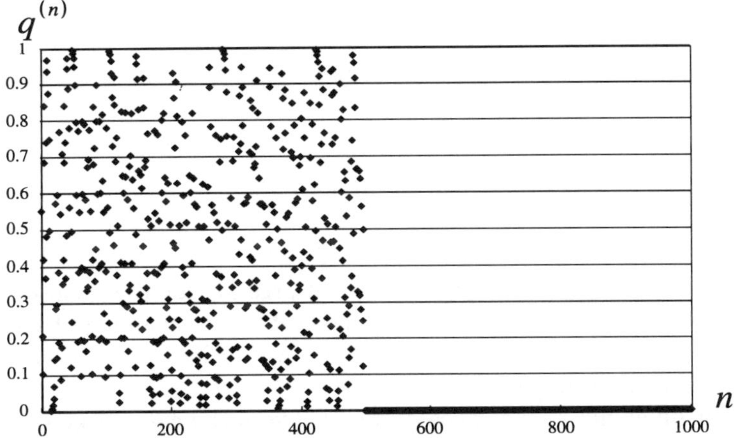

Fig.2. The distribution of the classical value $q^{(n)}$ for the case N=500

Fig.3 presents the change of the chaos degree for the case $N = 500$ up to the time $n = 1000$.

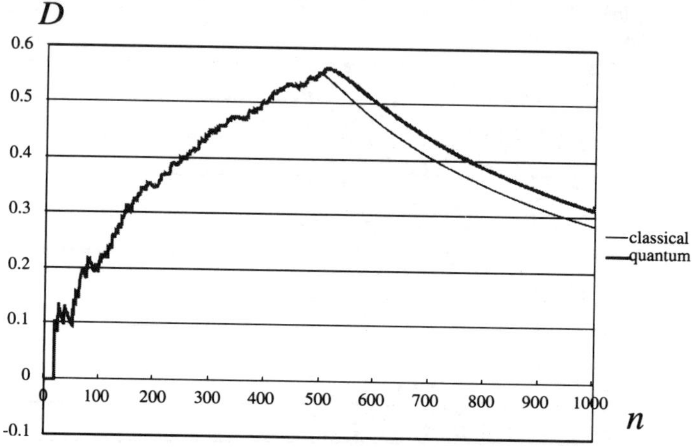

Fig.3. The change of the chaos degree for the case N=500 up to time n=1000

The correspondence between the chaos degree D_q for the quantum baker's map and the chaos degree D_c for the classical baker's map for a fixed N (500, here) is shown for the time less than $T = \log_2 \frac{1}{h} = \log_2 2^N = N$, and it is lost at the logarithtic time scale T.

The difference of the chaos degrees between the chaos degree D_q for the quantum baker's map and the chaos degree D_c for the classical baker's map for a fixed time n (1000, here) is displayed w.r.t. N in Fig.4.

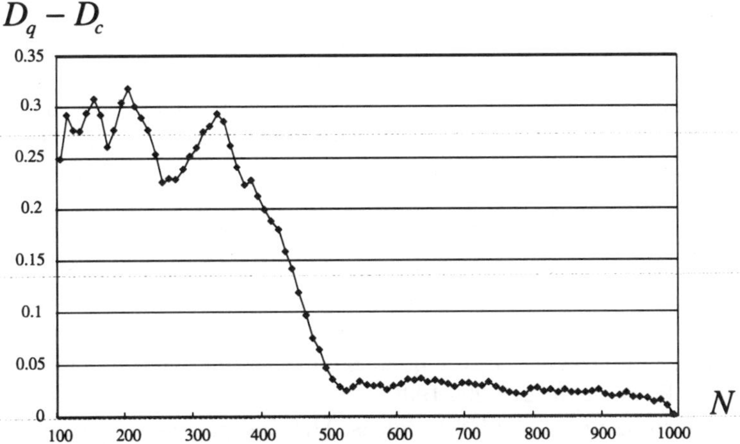

Fig.4. The difference of the chaos degree between quantum and classical for the case $n=1000$

Thus we conclude that the dynamics of the mean value $r_N^{(n)}$ reduces the classical dynamics $q^{(n)}$ in the q direction in the classical limit $N \to \infty \,(h \to 0)$.

References

1. D.V.Anosov and V.I.Arnold (eds.), *Dynamical Systems*, VINITI, Moscow, 1996.
2. V.M.Alekseev and M.N.Yakobson, Symbolic dynamics and hyperbolic dynamic systems, Phys. Reports, **75**, 287-325, 1981.
3. I.Y. Arefeva, P.B. Medvedev, O.A. Rytchkov and I.V. Volovich, Chaos in m(atrix) theory, Chaos, Solitons & Fractals, **10**, No.2-3, 213–223, 1999.
4. N.L.Balazs and A.Voros, The quantized baker's transformation, Ann. Phys., **190**, 1-31, 1989.
5. M.V.Berry, *Some quantum-to classical asymptotics*, Les Houches Summer School "chaos and quantum physics", Edits. Giannoni, M.J. Voros, A. and Justi, Zinn, North-Holland, Amsterdam, 1991.
6. T.Brun and R.Schack, Realizing thequantum baker's map on an NMR quantum computer, Phys. Rev. A, **59**, 2649-2658 , 1999.
7. G.Casati and B.V.Chirikov (eds.), *Quantum Chaos: between Order and Disorder*, Cambridge Univ. Press, Cambridge, 1995.
8. M.C.Gutzwiller, *Chaos in classical and Quantum Mechanics*, Springer, Berlin, 1990.
9. F.M.Dittes, E.Doron and U.Smilansky, Long-time behavior of the semi-

classical baker's map, Phys. Rev. E, **49**, R963- R966,1994.

10. K.Hepp, The classical limit for quantum mechanical correlation functions, Commun.Math.Phys., **35**, 265-277, 1974.

11. J.H.Hannay, J.P.Keating and A.M.Ozorio de Almeida, Optical realization of the baker's transformation, Nonlinearity, **7**, 1327-1342, 1994.

12. R.S.Ingarden, A.Kossakowski and M.Ohya, Information Dynamics and Open Systems, Kluwer Academic Publishers, 1997.

13. K.Inoue, M.Ohya and K.Sato, Application of chaos degree to some dynamical systems, Chaos, Soliton and Fractals, **11**, 1377-1385, 2000.

14. L.Kaplan and E.J.Heller, Overcoming the wall in the semiclassical baker's map, Phys. Rev. Lett., **76**, 1453-1456, 1996.

15. A.Lakshminarayan, On the quantum baker's map and its unusual traces, Ann. Phys., **239**, 272-295,1995.

16. A.Lakshminarayan and N.L.Balazs, The classical and quantum mechanics of Lazy baker maps, Ann. Phys., **226**, 350-373, 1993.

17. M.G.E. da Luz and A.M.Ozorio de Almeida, Path integral for the quantum baker's map, Nonlinearity, **8**, 43-64, 1995.

18. P.W.O'Connor and S.Tomsovic, The unusual nature of the quantum baker's transformation, Ann. Phys., **207**, 218-264, 1991.

19. M.Ohya, Complexities and their applications to characterization of chaos, International Journal of Theoretical Physics, **37**, No.1, 495-505, 1998.

20. A.M.Ozorio de Almeida and M.Saraceno, Periodic orbit theory for the quantized baker's map, Ann. Phys., **210**, 1-15, 1991.

21. M.Saraceno, Classical structures in the quantized baker transformation, Ann. Phys., **199**, 37-60, 1990.

22. M.Saraceno and A.Voros, Towards a semiclassical theory of the quantum baker's map. Physica D, **79**, 206-268, 1994.

23. R.Schack, Using a quantum computer to investigate quantum chaos, Phys. Rev. A, **57**, 1634-1635 ,1998.

24. R.Schack and C.M.Caves, Hypersensitivity to perturbations in the quantum baker's map, Phys. Rev. Lett., **71**, 525-528, 1993.

25. R.Schack and C.M.Caves, Information-theoretic characterization of quantum chaos, Phys. Rev. E., **53**, 3257-3270, 1996.

26. R.Schack and C.M.Caves, Shifts on a finite qubit string: A class of quantum baker's maps, Applicable Algebra in Engineering, Communication and Computing, **AAECC 10**, 305–310, 2000.

27. Ya.G.Sinai, *Introduction to Ergodic Theory*, Fasis, Moscow, 1996.

28. A. N. Soklakov and R. Schack, Classical limit in terms of symbolic dynamics for the quantum baker's map, e-print qunatu-ph/9908040.

29. G.M.Zaslavskii, *Stochasticity of Dynamical Systems*, Nauka, Moscow, 1984.

30. W.H.Zurek, Pointer basis of quantum apparatus: Into what mixture does the wave packet collapse?, Phys. Rev. D, **24**, 1516-1527, 1981.

Quantum Information IV (pp. 103–114)
Eds. T. Hida and K. Saitô
© 2002 World Scientific Publishing Co.

POISSON NOISE ANALYSIS BASED ON THE LÉVY LAPLACIAN

ATSUSHI ISHIKAWA

Graduate School of Mathematics
Meijo University
Nagoya 468-8502, Japan

KIMIAKI SAITÔ

Department of Information Sciences
Meijo University
Nagoya 468-8502, Japan

ALLANUS H. TSOI

Department of Mathematics
202 Mathematical Sciences Building
University of Missouri
Columbia, MO 65211, USA

In this paper we discuss the Lévy Laplacian acting on the space of Poisson noise functionals. In particular we introduce a semigroup generated by the Laplacian and give a stochastic expression of the semigroup.

Mathematics Subject Classification (2000): 60H40

1   Introduction

The Lévy Laplacian was introduced by P. Lévy [12] and was applied to the theory of generalized white noise functionals initiated by T. Hida [2]. This Laplacian has been studied by several authors within the framework of Gaussian white noise analysis (e.g. Refs. 1, 9, 10, 11, 13, 14–16). On the other hand the Poisson noise analysis has been also discussed in Refs. 3–8, 17, 18 and others.

In the previous paper [17], we introduced a domain of the Lévy Laplacian to get the self-adjointness of the Laplacian on the domain.

In this paper we introduce a domain consisting of Poisson noise functionals which are eigenfunctions of the Laplacian. The domain is larger than the one given in Ref.17. We define an equi-continuous semigroup of class (C_0) generated by the Laplacian on this domain and give also a stochastic expression of the semigroup acting on the domain.

The paper is organized as follows. In Section 2 we summarize some basic definitions and results in the Poisson noise analysis. In Section 3 we give the definition of the Lévy Laplacian acting on Poisson noise functionals and also give eigenfunctions of the Laplacian. In Section 4 we introduce Hilbert spaces $\mathbf{E}_{N,p}$ depending on the parameters $N \in \mathbf{N}$ and $p \in \mathbf{R}$ related to weights of norms. The spaces consist of eigenfunctions in Section 3 and so these are suitable to act as domains of the Laplacian. For $p \in \mathbf{R}$, the following inclusion relations hold:

$$\cdots \subset \mathbf{E}_{N+1,p} \subset \mathbf{E}_{N,p} \subset \cdots \subset \mathbf{E}_{1,p} \subset \mathbf{E}_{0,p} \subset \cdots \subset \mathbf{E}_{-N,p} \subset \mathbf{E}_{-N-1,p} \subset \cdots.$$

The Lévy Laplacian becomes a self-adjoint operator densely defined on $\mathbf{E}_{N,p}$ for each $N \in \mathbf{N}$ and $p \in \mathbf{R}$, and an equi-continuous semigroup of class (C_0) generated by the Laplacian is defined on the inductive limit space $\mathbf{E}_{-\infty,p}$ of $\mathbf{E}_{N,p}$, $N \in \mathbf{N}, p \in \mathbf{R}$. In the last section we give stochastic expressions of the semigroup generated by the Laplacian on the image space $\mathcal{U}[\mathbf{E}_{\infty,p}]$ of the projective limit space $\mathbf{E}_{\infty,p}$ of $\mathbf{E}_{N,p}$, $N \in \mathbf{N}, p \in \mathbf{R}$, by the $\mathcal{U}$-transform and the semigroup generated by the Laplacian on the space $\mathbf{E}_{\infty,p}$, respectively.

2 Preliminaries

In this section we summarize the basic definitions and results in Poisson noise calculus following Refs. 3 and 5 (see also Refs. 4 and 6.)

We start with the following Gel'fand triple

$$E \equiv \mathcal{S}(\mathbf{R}) \subset L^2(\mathbf{R}) \subset E^* \equiv \mathcal{S}'(\mathbf{R}),$$

where $\mathcal{S}(\mathbf{R})$ is the space of rapidly decreasing functions defined on $\mathbf{R}$, $L^2(\mathbf{R})$ is the Hilbert space consisting of square-integrable functions on $\mathbf{R}$ with norm $|\cdot|_0$, and $\mathcal{S}'(\mathbf{R})$ is the dual space of $\mathcal{S}(\mathbf{R})$, i.e. the space of tempered distributions. Let $A = -(d/du)^2 + u^2 + 1$. Then A is a densely defined self-adjoint operator on $L^2(\mathbf{R})$ and there exists an orthonormal basis $\{e_k; k = 0, 1, 2, \ldots\} \subset E$ for $L^2(\mathbf{R})$ satisfying $Ae_k = 2(k+1)e_k$, $k = 0, 1, 2, \ldots$. For each $p \in \mathbf{R}$, we define $|\xi|_p = |A^p\xi|_0$ and let $E_p = \{\xi \in L^2(\mathbf{R}); |\xi|_p < \infty\}$ for each $p \geq 0$ and let E_p be the completion of $L^2(\mathbf{R})$ with norm $|\cdot|_p$ for each $p < 0$. We introduce the projective limit topology and the inductive limit topology of spaces $E_p, p \in \mathbf{R}$ to E and E^*, respectively.

The Bochner-Minlos Theorem admits the existence of a probability measure μ on E^* whose characteristic functional is given by

$$\int_{E^*} \exp\{i\langle x, \xi\rangle\}d\mu(x) = \exp\left[\int_{\mathbf{R}} \left(e^{i\xi(u)} - 1\right) du\right], \ \xi \in E,$$

where $\langle \cdot, \cdot \rangle$ is the canonical bilinear form between E^* and E.

Let $(L^2) \equiv L^2(E^*, \mu)$ be a Hilbert space consisting of complex-valued square-integrable functionals defined on a Poisson noise probability space (E^*, μ). Take a sequence $\{\eta_n\}_{n=1}^{\infty} \subset E$ converging to $1_{(0,t]}$ in $L^2(\mathbf{R})$ for each $t \geq 0$ and to $-1_{(t,0]}$ in $L^2(\mathbf{R})$ for each $t < 0$. Then the sequence $\{\langle \cdot, \eta_n \rangle\}_{n=1}^{\infty}$ converges to a Poisson noise functional $N(t) \equiv N.(t)$ in (L^2) and $N(t)$ becomes a standard Poisson process defined on (E^*, μ).

The $\mathcal{U}$-transform is defined by

$$\mathcal{U}\varphi(\xi) = C(\xi)^{-1} \int_{E^*} e^{i\langle x, \xi \rangle} \varphi(x) d\mu(x), \quad \xi \in E,$$

for each $\varphi \in (L^2)$, where $C(\xi) = \exp\left[\int_{\mathbf{R}} \left(e^{i\xi(u)} - 1 \right) du \right]$, $\xi \in E$. If we introduce an inner product

$$(\mathcal{U}\varphi, \mathcal{U}\psi)_{\mathbf{U}} = (\varphi, \psi)_{(L^2)}, \quad \text{for} \quad \varphi, \psi \in (L^2),$$

to $\mathbf{U} = \mathcal{U}[(L^2)]$, then $\mathbf{U}$ is a reproducing kernel Hilbert space with kernel

$$K(\eta, \zeta) = e^{\langle e^{i\eta} - 1, e^{i\zeta} - 1 \rangle}, \quad \eta, \zeta \in E.$$

The $\mathcal{U}$-transform is an isomorphism from (L^2) onto $\mathbf{U}$.

Put $Q(t) = N(t) - t$. Then $Q(t)$ is the compensated Poisson process and the space (L^2) has the Wiener-Itô decomposition:

$$(L^2) = \bigoplus_{n=0}^{\infty} H_n,$$

where H_n is a closed subspace of (L^2) given by

$$H_n = \left\{ \mathbf{I}_n(f) = \int_{\mathbf{R}^n} f(\mathbf{u}) dQ(u_1) \cdots dQ(u_n); \ f \in L_{\mathbf{C}}^2(\mathbf{R})^{\hat{\otimes} n} \right\},$$

where $\mathbf{u} = (u_1, \ldots, u_n)$ and $L_{\mathbf{C}}^2(\mathbf{R})^{\hat{\otimes} n}$ is the n-fold symmetric tensor product of the complexification $L_{\mathbf{C}}^2(\mathbf{R})$ of $L^2(\mathbf{R})$. The (L^2)-norm $\|\varphi\|_0$ of $\varphi = \sum_{n=0}^{\infty} \mathbf{I}_n(f_n) \in (L^2)$ is given by

$$\|\varphi\|_0 = \left(\sum_{n=0}^{\infty} n! |f_n|_{L_{\mathbf{C}}^2(\mathbf{R})^{\otimes n}}^2 \right)^{1/2}.$$

The $\mathcal{U}$-transform $\mathcal{U}\varphi$ of $\varphi = \sum_{n=0}^{\infty} \mathbf{I}_n(f_n) \in (L^2)$ is given by

$$\mathcal{U}\varphi(\xi) = \sum_{n=0}^{\infty} \int_{\mathbf{R}^n} f_n(\mathbf{u}) \prod_{\nu=1}^{n} (e^{i\xi(u_\nu)} - 1) d\mathbf{u}.$$

The second quantization operator $\Gamma(A)$ of A is densely defined on (L^2) by

$$\Gamma(A)\varphi = \sum_{n=0}^{\infty} \mathbf{I}_n(A^{\otimes n} f_n)$$

for $\varphi = \sum_{n=0}^{\infty} \mathbf{I}_n(f_n)$. For $p \in \mathbf{R}$, let $\|\varphi\|_p = \|\Gamma(A)^p \varphi\|_0$. If $p \geq 0$, let $(E)_p$ be the domain of $\Gamma(A)^p$. If $p < 0$, let $(E)_p$ be the completion of (L^2) with respect to the norm $\|\cdot\|_p$. Then $(E)_p, p \in \mathbf{R}$, is a Hilbert space with th norm $\|\cdot\|_p$. It is easy to see that for $p > 0$, the dual space $(E)_p^*$ of $(E)_p$ is given by $(E)_{-p}$. Moreover, for any $p \in \mathbf{R}$, we have the decomposition

$$(E)_p = \bigoplus_{n=0}^{\infty} H_n^{(p)},$$

where $H_n^{(p)}$ is the completion of $\{\mathbf{I}_n(f); f \in E_{\mathbf{C}}^{\hat{\otimes} n}\}$ with respect to $\|\cdot\|_p$. Here $E_{\mathbf{C}}^{\hat{\otimes} n}$ is the n-fold symmetric tensor product of the complexification $E_{\mathbf{C}}$ of E. We also have $H_n^{(p)} = \{\mathbf{I}_n(f); f \in E_{\mathbf{C},p}^{\hat{\otimes} n}\}$ for any $p \in \mathbf{R}$, where $E_{\mathbf{C},p}^{\hat{\otimes} n}$ is also the n-fold symmetric tensor product of the complexification $E_{\mathbf{C},p}$ of E_p. The norm $\|\varphi\|_p$ of $\varphi = \sum_{n=0}^{\infty} \mathbf{I}_n(f_n) \in (E)_p$ is given by

$$\|\varphi\|_p = \left(\sum_{n=0}^{\infty} n! |f_n|_p^2 \right)^{1/2}, \quad f_n \in E_{\mathbf{C},p}^{\hat{\otimes} n},$$

where the norm of $E_{\mathbf{C},p}^{\otimes n}$ is denoted also by $|\cdot|_p$.

The projective limit space (E) of spaces $(E)_p, p \in \mathbf{R}$ is a nuclear space. The inductive limit space $(E)^*$ of spaces $(E)_p, p \in \mathbf{R}$ is nothing but the dual space of (E). The space $(E)^*$ is called the space of *generalized Poisson noise functionals*. We denote by $\ll \cdot, \cdot \gg$ the canonical bilinear form on $(E)^* \times (E)$. Then we have

$$\ll \Phi, \varphi \gg = \sum_{n=0}^{\infty} n! \langle F_n, f_n \rangle$$

for any $\Phi = \sum_{n=0}^{\infty} \mathbf{I}_n(F_n) \in (E)^*$ and $\varphi = \sum_{n=0}^{\infty} \mathbf{I}_n(f_n) \in (E)$, where the canonical bilinear form on $(E_{\mathbf{C}}^{\otimes n})^* \times E_{\mathbf{C}}^{\otimes n}$ is denoted also by $\langle \cdot, \cdot \rangle$.

Since $\phi_\xi(\cdot) \equiv C(\xi)^{-1} \exp\{i\langle \cdot, \xi \rangle\} \in (E)$, the $\mathcal{U}$-*transform* can be extended to a continuous linear operator on $(E)^*$ by

$$\mathcal{U}\Phi(\xi) = \ll \Phi, \phi_\xi \gg, \quad \xi \in E.$$

We have the following characterization theorem of the $\mathcal{U}$-transform.

Theorem 2.1. *Let F be a complex-valued function defined on $E_{\mathbf{C}}$. Then F is the $\mathcal{U}$-transform of some generalized functional if and only if there exists a complex-valued function G such that*

1) for any ξ and η in $E_{\mathbf{C}}$, the function $G(z\xi + \eta)$ is an entire function of $z \in \mathbf{C}$,

2) there exist nonnegative constants $K, a,$ and p such that

$$|G(\xi)| \le K \exp\left[a|\xi|_p^2\right], \quad \forall \xi \in E_{\mathbf{C}},$$

3) $F(\xi) = G(e^{i\xi} - 1)$ for all $\xi \in E$.

3 The Lévy Laplacian acting on Poisson noise functionals

Let Φ be in $(E)^*$. Then for the $\mathcal{U}$-transform $\mathcal{U}\Phi$, there exists a complex-valued function G satisfying conditions 1), 2), 3) in Theorem 2.1. In particular, for any $\xi \in E$, the second variation of $\mathcal{U}\Phi$ is given by

$$\mathcal{U}\Phi''(\xi)(\eta, \zeta) = G''(e^{i\xi} - 1)(i\eta e^{i\xi}, i\zeta e^{i\xi}) + G'(e^{i\xi} - 1)(i\zeta e^{i\xi}), \quad \eta, \zeta \in E.$$

Fix a finite interval T of $\mathbf{R}$. Take an orthonormal basis $\{\zeta_n\}_{n=0}^{\infty} \subset E$ for $L^2(T)$ satisfying the equally dense and uniform boundedness condition (see Refs. 9, 12, 13). Let $\mathcal{D}_L$ denote the set of all $\Phi \in (E)^*$ such that the limit

$$\widetilde{\Delta}_L \mathcal{U}\Phi(\xi) = \lim_{N \to \infty} \frac{1}{N} \sum_{n=0}^{N-1} (\mathcal{U}\Phi)''(\xi)(\zeta_n, \zeta_n)$$

exists for any $\xi \in E_{\mathbf{C}}$ and is in $\mathcal{U}[(E)^*]$. Then we consider the *Lévy Laplacian* Δ_L defined by

$$\Delta_L \Phi = \mathcal{U}^{-1} \widetilde{\Delta}_L \mathcal{U}\Phi$$

for $\Phi \in \mathcal{D}_L$. We denote a set of all functionals $\Phi \in \mathcal{D}_L$ such that $\mathcal{U}\Phi(\eta) = 0$ for all $\eta \in E$ with $\text{supp}(\eta) \subset T^c$ by $\mathcal{D}_L^T$.

For any $f \in E_{\mathbf{C}}^{\hat{\otimes} n}$ with $\text{supp}(f) \subset T^n$, a Poisson noise functional

$$\varphi = \int_{T^n} f(u_1, \ldots, u_n) dN(u_1) \cdots dN(u_n)$$

is in $\mathcal{D}_L^T$ and is equal to

$$\sum_{j=0}^{n} \frac{n!}{j!(n-j)!} \int_{T^n} f(\mathbf{u}) dQ(u_1) \cdots dQ(u_{n-j}) du_{n-j+1} \cdots du_n.$$

Hence, the $\mathcal{U}$-transform $\mathcal{U}\varphi$ of φ is given by

$$\mathcal{U}\varphi(\xi) = \int_{T^n} f(\mathbf{u})e^{i\xi(u_1)}\cdots e^{i\xi(u_n)}d\mathbf{u}.$$

We put

$$\mathbf{D}_n = \left\{ \int_{T^n} f(\mathbf{u})dN(u_1)\cdots dN(u_n); f \in E_{\mathbf{C}}^{\hat{\otimes}n}, \ \mathrm{supp}(f) \subset T^n \right\}$$

for each $n \in \mathbf{N}$.

Then for each $n \in \mathbf{N}$, $\mathbf{D}_n$ is a linear subspace of (E) and therefore for $p \in \mathbf{R}$, we can define a space $\overline{\mathbf{D}}_n^{(p)}$ by the completion of $\mathbf{D}_n$ in $(E)_p$ with respect to $\|\cdot\|_p$. Define $\overline{\mathbf{D}}_0^{(p)}$ by the complex field $\mathbf{C}$. Then for each $n \in \mathbf{N} \cup \{0\}$ and $p \in \mathbf{R}$, $\overline{\mathbf{D}}_n^{(p)}$ becomes a Hilbert space with the inner product of $(E)_p$. The Lévy Laplacian Δ_L becomes a linear operator defined on $\overline{\mathbf{D}}_n^{(p)}$ for each $n \in \mathbf{N} \cup \{0\}$ and $p \in \mathbf{R}$.

Theorem 3.1.[Ref.17] *For each $n \in \mathbf{N} \cup \{0\}, p \in \mathbf{R}$ and $\varphi \in \overline{\mathbf{D}}_n^{(p)}$, the equation*

$$\Delta_L \varphi = -\frac{n}{|T|}\varphi.$$

holds.

4 A semi-group generated by the Lévy Laplacian

If $\Phi \in (E)^*$ is expressed in the form $\sum_{n=0}^{\infty} \Phi_n$, $\Phi_n \in \overline{\mathbf{D}}_n^{(p)}$, $n = 0, 1, \ldots$, then the expression is uniquely determined in $(E)^*$. Then we can define the following spaces.

Let $\alpha_N(n) = \sum_{\ell=0}^{N} \left(\frac{n}{|T|}\right)^{2\ell}$ for each $N \in \mathbf{N} \cup \{0\}$ and $n \in \mathbf{N} \cup \{0\}$. Set

$$\mathbf{E}_{N,p} = \left\{ \sum_{n=0}^{\infty} \varphi_n; \sum_{n=0}^{\infty} \alpha_N(n)\|\varphi_n\|_p^2 < \infty, \varphi_n \in \overline{\mathbf{D}}_n^{(p)}, n = 0, 1, 2, \ldots \right\}$$

for each $N \in \mathbf{N}$ and $p \in \mathbf{R}$. Then for any $N \in \mathbf{N}$, $\mathbf{E}_{N,p}$ is in $(E)_p$ and is a Hilbert space with norm $\|\|\cdot\|\|_{N,p}$ given by

$$\|\|\varphi\|\|_{N,p} = \left(\sum_{n=0}^{\infty} \alpha_N(n)\|\varphi_n\|_p^2 \right)^{1/2}, \quad \varphi = \sum_{n=0}^{\infty} \varphi_n \in \mathbf{E}_{N,p}.$$

Put $\mathbf{E}_{\infty,p} = \bigcap_{N=1}^{\infty} \mathbf{E}_{N,p}$ with the projective limit topology.

For any $N \in \mathbf{N} \cup \{0\}$ and $p \in \mathbf{R}$, we define $\mathbf{E}_{-N,p}$ by the completion of $\mathbf{E}_{\infty,p}$ with respect to norm $||| \cdot |||_{-N,p}$ given by

$$|||\varphi|||_{-N,p} = \left(\sum_{n=0}^{\infty} \alpha_N(n)^{-1} \|\varphi_n\|_p^2 \right)^{1/2}, \quad \varphi = \sum_{n=0}^{\infty} \varphi_n \in \mathbf{E}_{\infty,p}.$$

Put $\mathbf{E}_{-\infty,p} = \bigcup_{N=1}^{\infty} \mathbf{E}_{-N,p}$ with the inductive limit topology. Then for $p \in \mathbf{R}$, we have the following inclusion relations:

$$\mathbf{E}_{\infty,p} \subset \cdots \subset \mathbf{E}_{N+1,p} \subset \mathbf{E}_{N,p} \subset \cdots \subset \mathbf{E}_{1,p} \subset \mathbf{E}_{0,p}$$

$$\subset \cdots \subset \mathbf{E}_{-N,p} \subset \mathbf{E}_{-N-1,p} \subset \cdots \subset \mathbf{E}_{-\infty,p}.$$

The space $\mathbf{E}_{\infty,p}$ includes $\overline{\mathbf{D}}_n^{(p)}$ for any $n \in \mathbf{N} \cup \{0\}$ and $p \in \mathbf{R}$. The Laplacian Δ_L can be extended to a continuous linear operator defined on $\mathbf{E}_{-\infty,p}$ into $\mathbf{E}_{-\infty,p}$, denoted by the same notation Δ_L, for each $p \in \mathbf{R}$.

For each $t \geq 0$ and $p \in \mathbf{R}$, we consider an operator G_t on $\mathbf{E}_{-\infty,p}$ defined by

$$G_t\varphi = \sum_{n=0}^{\infty} e^{-t\frac{n}{|T|}} \varphi_n$$

for $\varphi = \sum_{n=0}^{\infty} \varphi_n \in \mathbf{E}_{-\infty,p}$. Then we have the following:

Theorem 4.1. [cf. Ref. 16] *Let $p \in \mathbf{R}$. Then the family $\{G_t; t \geq 0\}$ is an equi-continuous semigroup of class (C_0) generated by Δ_L as a continuous linear operator defined on $\mathbf{E}_{-\infty,p}$.*

Proof: Let $p \in \mathbf{R}$. For any $\Phi \in \mathbf{E}_{-\infty,p}$, there exists $N \in \mathbf{N}$ such that $\Phi \in \mathbf{E}_{-N,p}$. Then, for any $t \geq 0$ the norm $|||G_t\Phi|||_{-N,p}$ for $\Phi = \sum_{n=0}^{\infty} \Phi_n \in \mathbf{E}_{-\infty,p}$ is estimated as follows:

$$|||G_t\Phi|||^2_{-N,p} = \sum_{n=0}^{\infty} \alpha_N(n)^{-1} \|e^{-t\frac{n}{|T|}} \Phi_n\|_p^2$$

$$\leq \sum_{n=0}^{\infty} \alpha_N(n)^{-1} \|\Phi_n\|_p^2$$

$$= |||\Phi|||^2_{-N,p}.$$

110

Hence the family $\{G_t; t \geq 0\}$ is equi-continuous in t. It is easily checked that $G_0 = I, G_t G_s = G_{t+s}$ for each $t, s \geq 0$. We can also estimate that

$$|||G_t \Phi - G_{t_0} \Phi|||^2_{-N,p} = \sum_{n=0}^{\infty} \alpha_N(n)^{-1} \left| e^{-t\frac{n}{|T|}} - e^{-t_0 \frac{n}{|T|}} \right|^2 ||\Phi_n||_p^2$$

$$\leq 4 \sum_{n=0}^{\infty} \alpha_N(n)^{-1} ||\Phi_n||_p^2$$

$$= 4|||\Phi|||^2_{-N,p} < \infty$$

for each $t, t_0 \geq 0$. Therefore, by the Lebesgue convergence theorem, we get that

$$\lim_{t \to t_0} G_t \Phi = G_{t_0} \Phi \quad \text{in} \quad \mathbf{E}_{-\infty, p}$$

for each $t_0 \geq 0$ and $\Phi \in \mathbf{E}_{-\infty, p}$. Thus the family $\{G_t; t \geq 0\}$ is an equi-continuous semigroup of class (C_0). We next prove that the infinitesimal generator of the semigroup is given by Δ_L. For any $\Phi = \sum_{n=0}^{\infty} \Phi_n \in \mathbf{E}_{-\infty, p}$, there exists $N \in \mathbf{N}$ such that $|||\Phi|||_{-N,p} < \infty$. We note that

$$\left|\left|\left| \frac{G_t \Phi - \Phi}{t} - \Delta_L \Phi \right|\right|\right|^2_{-(N+1),p} = \sum_{n=0}^{\infty} \alpha_{N+1}(n)^{-1} \left|\left| \frac{e^{-t\frac{n}{|T|}} - 1}{t} \Phi_n - \frac{n}{|T|} \Phi_n \right|\right|^2_p.$$

By the mean value theorem, for any $t > 0$ there exists a constant $\theta \in (0, 1)$ such that

$$\left| \frac{e^{-t\frac{n}{|T|}} - 1}{t} \right| = \frac{n}{|T|} e^{-t\theta \frac{n}{|T|}} < \frac{n}{|T|}.$$

Therefore we can estimate the following term:

$$\alpha_{N+1}(n)^{-1} \left|\left| \frac{e^{-t\frac{n}{|T|}} - 1}{t} \Phi_n - \frac{n}{|T|} \Phi_n \right|\right|^2_p = \alpha_{N+1}(n)^{-1} \left| \frac{e^{-t\frac{n}{|T|}} - 1}{t} - \frac{n}{|T|} \right|^2 ||\Phi_n||_p^2$$

$$\leq 4\alpha_N(n)^{-1} ||\Phi_n||_p^2.$$

By

$$\lim_{t \to 0} \left| \frac{e^{-t\frac{n}{|T|}} - 1}{t} - \frac{n}{|T|} \right| = 0$$

and the Lebesgue convergence theorem, we obtain

$$\lim_{t \to 0} \left\|\left\| \frac{G_t \Phi - \Phi}{t} - \Delta_L \Phi \right\|\right\|_{-(N+1),p}^2 = 0.$$

Thus the proof is completed. $\qquad\square$

5 Stochastic expressions of the semigroup generated by the Lévy Laplacian

Let $\{X_t; t \geq 0\}$ be a Cauchy process with the characteristic function of X_t given by

$$E[e^{izX_t}] = e^{-t|z|}, \quad z \in \mathbf{R}.$$

Take a smooth function $\eta_T \in E$ with $\eta_T = \frac{1}{|T|}$ on T. Let $p \in \mathbf{R}$ and set $\widetilde{G}_t = \mathcal{U} G_t \mathcal{U}^{-1}$ on $\mathcal{U}[\mathbf{E}_{\infty,p}]$ with the topology induced from $\mathbf{E}_{\infty,p}$ by the $\mathcal{U}$-transform. Then by Theorem 4.1, $\{\widetilde{G}_t; t \geq 0\}$ is an equi-continuous semigroup of class (C_0) generated by the operator $\widetilde{\Delta}_L$.

Let $\{\mathbf{X}_t; t \geq 0\}$ be an E-valued stochastic process given by $\mathbf{X}_t = \xi + X_t \eta_T$, $\xi \in E$. Then we have the following a stochastic expression of the operator $\widetilde{G}_t$.

Theorem 5.1. *Let p be a real number. Then we have*

$$\widetilde{G}_t F(\xi) = \mathrm{E}[F(\mathbf{X}_t)|\mathbf{X}_0 = \xi]$$

for all $F \in \mathcal{U}[\mathbf{E}_{\infty,p}]$.

Proof: Using the similar method in the proof of Theorem 7 in Ref. 16, we can obtain this theorem. $\qquad\square$

By Theorem 2.1 we can prove that for Φ and Ψ in $(E)^*$, there exists a unique generalized Poisson noise functional whose $\mathcal{U}$-transform is given by $(\mathcal{U}\Phi)(\mathcal{U}\Psi)$. We call this generalized functional in $(E)^*$ the *Wick product* of Φ and Ψ in $(E)^*$, denoted by $\Phi \diamond \Psi$, i.e.

$$\mathcal{U}(\Phi \diamond \Psi) = (\mathcal{U}\Phi)(\mathcal{U}\Psi).$$

Since for $x \in E^*$ and $\eta \in E$, the product $e^{i\eta}x$ is given by $e^{i\eta}x = x + (e^{i\eta} - 1)x$, we can define a continuous linear operator $M_{e^{i\eta}}$ on $\mathbf{E}_{\infty,p}$ characterized by

$$M_{e^{i\eta}}[\diamond_{j=1}^n \langle \cdot, f_j \rangle](x) = \diamond_{j=1}^n \langle e^{i\eta}x, f_j \rangle \quad n \in \mathbf{N} \cup \{0\}.$$

Here $\diamond_{j=1}^{n} \langle \cdot, f_j \rangle$ means $\langle \cdot, f_1 \rangle \diamond \cdots \diamond \langle \cdot, f_n \rangle$ and we note that it is equal to

$$\int_{T^n} f_1(u_1) \cdots f_n(u_n) dN_x(u_1) \cdots dN_x(u_n)$$

with $\mathrm{supp}(f_j) \subset T$, $j = 1, 2, \ldots, n$. Then we have the following.

Theorem 5.2. *Let p be a real number. Then for any $t \geq 0$ and $\varphi \in \mathbf{E}_{\infty,p}$, the equality*

$$G_t \varphi = \mathrm{E}[M_{e^{iX_t \eta_T}} \varphi]$$

holds.

Proof: Put $\varphi = \int_{T^n} f_1(u_1) \cdots f_n(u_n) dN(u_1) \cdots dN(u_n)$ in $\mathbf{E}_{\infty,p}$ with $\mathrm{supp}(f_j) \subset T$, $j = 1, 2, \ldots, n$. Then we have

$$\begin{aligned}
\mathrm{E}[M_{e^{iX_t \eta_T}} \varphi(x)] &= \mathrm{E}[\diamond_{j=1}^{n} \langle e^{iX_t \eta_T} x, f_j \rangle] \\
&= \mathrm{E}[e^{i \frac{n}{|T|} X_t} \diamond_{j=1}^{n} \langle x, f_j \rangle] \\
&= e^{-t \frac{n}{|T|}} \varphi(x) = G_t \varphi(x).
\end{aligned}$$

Let $\varphi = \sum_{n=0}^{\infty} \varphi_n \in \mathbf{E}_{\infty,p}$. Then for any $n \in \mathbf{N} \cup \{0\}$, φ_n is expressed in the following form:

$$\varphi_n = \lim_{\ell \to \infty} \sum_{k_1, \ldots, k_n = 1}^{N_\ell} a_{k_1, \ldots, k_n}^{[\ell]} \diamond_{j=1}^{n} \langle \cdot, f_{n,k_j}^{[\ell]} \rangle, \quad \text{in } \mathbf{E}_{\infty,p},$$

where $(a_{k_1, \ldots, k_n}^{[\ell]})_{k_1, \ldots, k_n, \ell}$ is a sequence of complex numbers and $(f_{n,k}^{[\ell]})_{k,\ell}$ is a sequence of functions in $E_{\mathbf{C}}^{\hat{\otimes} n}$. Hence for any $N \in \mathbf{N} \cup \{0\}$, we have

$$\sum_{n=0}^{\infty} \mathrm{E}[|||M_{e^{iX_t \eta_T}} \varphi_n|||_{N,p}^2]$$

$$= \sum_{n=0}^{\infty} \mathrm{E}\left[\lim_{\ell \to \infty} \left|\left|\left|\sum_{k_1, \ldots, k_n = 1}^{N_\ell} a_{k_1, \ldots, k_n}^{[\ell]} \diamond_{j=1}^{n} \langle e^{iX_t \eta_T}(\cdot), f_{n,k_j}^{[\ell]} \rangle \right|\right|\right|_{N,p}^2\right]$$

$$= \sum_{n=0}^{\infty} \lim_{\ell \to \infty} \left|\left|\left| \sum_{k_1, \ldots, k_n = 1}^{N_\ell} a_{k_1, \ldots, k_n}^{[\ell]} \diamond_{j=1}^{n} \langle \cdot, f_{n,k_j}^{[\ell]} \rangle \right|\right|\right|_{N,p}^2$$

$$= \sum_{n=0}^{\infty} |||\varphi_n|||_{N,p}^2 < \infty.$$

By the Schwarz inequality, $\sum_{n=0}^{\infty} E[|||M_{e^{iX_t\eta_T}}\varphi_n|||_{N,p}] < \infty$ for all $N \in \mathbf{N} \cup \{0\}$. Therefore by the continuity of G_t we get that

$$
\begin{aligned}
E[M_{e^{iX_t\eta_T}}\varphi] &= \sum_{n=0}^{\infty} E[M_{e^{iX_t\eta_T}}\varphi_n] \\
&= \sum_{n=0}^{\infty} G_t\varphi_n \\
&= G_t\varphi.
\end{aligned}
$$

Thus the proof is completed. $\qquad\square$

Acknowledgments

This work was written based on the previous works K.S. and A.H.T. [17] and also K.S. [15] [16]. This work was supported in part by the Joint Research Project " Quantum Information Theoretical Approach to Life Science " for the Academic Frontier in Science and was also supported in part by JSPS-PAN Joint Research Project "Infinite Dimensional Harmonic Analysis" promoted by the Ministry of Education in Japan. The authors are grateful for the support.

References

1. D.M. Chung, U.C. Ji and K. Saitô: Cauchy problems associated with the Lévy Laplacian in white noise analysis, to appear in Journal of Infinite Dimensional Analysis, Quantum Probability and Related Topics, Vol. 2, No. 1 (1999).
2. T. Hida; Analysis of Brownian Functionals, Carleton Math. Lecture Notes, No.13, Carleton University, Ottawa (1975).
3. T. Hida and N. Ikeda; Analysis on Hilbert space with reproducing kernel arising from multiple Wiener integral, in: Proc. 5th Berkeley Symp. on Math. Stat. and Prob. Vol. II, Part 1 (1967) 117 -143.
4. K. Itô; Spectral type of the shift transformation of differential processes with stationary increments, Trans. Amer. Math. Soc. 81 (1956) 253-263.
5. Y. Ito; Generalized Poisson Functionals, Probab. Th. Rel. Fields 77 (1988) 1-28.
6. Y.Ito and I. Kubo; Calculus on Gaussian and Poisson white noises, Nagoya Math. J. , Vol. 111 (1988) 41-84.

114

7. Y.G. Kondratiev, J.L. Da Silva, L. Streit and G.H. Us; Analysis on Poisson and Gamma spaces, Infinite Dim. Anal. Quantum Probab. Rel. Topics **1** (1998) 91-117.

8. Y.G. Kondratiev, L. Streit, W. Westerkamp and J.-A. Yan; Generalized functions in infinite dimensional analysis, IIAS Report, No. 1995-002 (1995).

9. H.-H. Kuo; White noise distribution theory, CRC Press, Boca Raton (1996).

10. H.-H. Kuo, N. Obata and K. Saitô; Lévy Laplacian of generalized functions on a nuclear space, J. Funct. Anal. 94 (1990) 74-92.

11. H.-H. Kuo, N. Obata and K. Saitô; Diagonalization of the Lévy Laplacian and Related Stable Processes, submitted to Infinite Dim. Anal. Quantum Probab. Rel. Topics (2001).

12. P. Lévy; Lecons d'analyse fonctionnelle, Gauthier-Villars, Paris (1922).

13. N. Obata; White Noise Calculus and Fock Space, Lecture Notes in Mathematics 1577, Springer-Verlag (1994).

14. K. Saitô; A C_0-group generated by the Lévy Laplacian II, Infinite Dimensional Analysis, Quantum Probability and Related Topics **Vol. 1**, No. 3 (1998) 425-437.

15. K. Saitô; A stochastic process generated by the Lévy Laplacian, *Acta Applicandae Mathematicae* **63** (2000) 363-373.

16. K. Saitô; The Lévy Laplacian and stable processes, *Chaos, Solitons and Fractals* **12** (2001) 2865-2872.

17. K. Saitô, K., A. H. Tsoi; The Lévy Laplacian acting on Poisson noise functionals, *Infinite Dimensional Analysis, Quantum Probability and Related Topics* **Vol. 2** (1999), 503–510.

18. A.H. Tsoi; L-transform, Normal Functionals and Lévy Laplacian in Poisson Noise Analysis, Preprint (2000).

Quantum Information IV (pp. 115–126)
Eds. T. Hida and K. Saitô
© 2002 World Scientific Publishing Co.

A HAUSDORFF-YOUNG INEQUALITY FOR WHITE NOISE ANALYSIS

HUI-HSIUNG KUO

Department of Mathematics, Louisiana State University
Baton Rouge, LA 70803, U.S.A.

KIMIAKI SAITÔ

Department of Information Sciences, Meijo University
Nagoya 468-8502, Japan

AUREL STAN

Department of Mathematics, Northwestern University
Evanston, IL 60208-2730, U.S.A.

A Hausdorff-Young inequality and Young inequality are proven for White Noise Analysis. The Fourier transform from the classical case is replaced by the S-transform or the Segal-Bargmann transform.

1   INTRODUCTION

The classical Hausdorff-Young inequality [10] says the following:
Theorem 1 *Let $n \in \mathbb{N}$. If $1 \le p \le 2$ and $2 \le q \le \infty$ such that $\frac{1}{p} + \frac{1}{q} = 1$, then the Fourier transform is a bounded linear map from $L^p(\mathbb{R}^n)$ into $L^q(\mathbb{R}^n)$.*
This theorem is related to the following Young inequality [10]:
Theorem 2 *Let $n \in \mathbb{N}$. Let $1 \le p, q, r \le \infty$ such that $\frac{1}{p} + \frac{1}{q} = 1 + \frac{1}{r}$. If $f \in L^p(\mathbb{R}^n)$ and $g \in L^q(\mathbb{R}^n)$, then their convolution product $f * g$ is in $L^r(\mathbb{R}^n)$ and the following inequality holds:*

$$\| f * g \|_r \le \| f \|_p \| g \|_q .$$

We will replace the classical Fourier transform by the S-transform and the convolution product by the Wick product to obtain similar results for White Noise Analysis.

2   BACKGROUND

We present below the basic background from White Noise Analysis. For more details see [7] and [9].

(a) Gel'fand triples

Let E be a real separable Hilbert space. We denote the norm of E by $|\cdot|_0$. Let A be a densely defined self-adjoint operator on E, having a countable spectrum and whose eigenvalues $\{\lambda_n\}_{n\geq 1}$ satisfy the following conditions:

- $1 < \lambda_1 \leq \lambda_2 \leq \lambda_3 \leq \cdots$.

- $\sum_{n=1}^{\infty} \lambda_n^{-2} < \infty$.

For any $p \geq 0$, let $\mathcal{E}_p = \{f \in E \mid |A^p f|_0 < \infty\}$. For any $f \in \mathcal{E}_p$, we define $|f|_p := |A^p f|_0$. We observe that $\mathcal{E}_p$ is a Hilbert space with the norm $|\cdot|_p$. The first condition implies that $\mathcal{E}_q \subset \mathcal{E}_p$, for any $q \geq p$. The second condition on the eigenvalues above is equivalent to the fact that A is a Hilbert-Schmidt operator from E into itself. Thus, for any $p \geq 0$, the inclusion map $i : \mathcal{E}_{p+1} \longrightarrow \mathcal{E}_p$ is a Hilbert-Schmidt operator.

Let $\mathcal{E} = \bigcap_{p \geq 0} \mathcal{E}_p$. The space $\mathcal{E}$ equipped with the topology given by the family $\{|\cdot|_p\}_{p \geq 0}$ of norms is a nuclear space. We denote by $\mathcal{E}'$ the dual of the nuclear space $\mathcal{E}$ and by $\mathcal{E}'_p$ the dual of the Hilbert space $\mathcal{E}_p$, $p \geq 0$. For any $p \geq 0$, the space $\mathcal{E}'_p$ is isomorphic to the space $\mathcal{E}_{-p}$, which is the completion of the initial Hilbert space E with respect to the norm $|\cdot|_{-p}$ defined in the following way: if $f \in E$, then $|f|_{-p} := |A^{-p} f|_0$. In this way the dual of $E_0 = E$ is isomorphic to $E_0 = E$. Therefore we have identified the dual of the initial Hilbert space E with E by Riesz representation theorem. If $0 < p < q$, then $\mathcal{E}_{-p} \subset \mathcal{E}_{-q}$. It turns out that $\mathcal{E}' = \bigcup_{p \geq 0} \mathcal{E}_{-p}$. The space $\mathcal{E}'$ is equipped with the inductive limit topology of the topologies of the increasing family of Hilbert spaces $\mathcal{E}_{-p}$. The duality between $\mathcal{E}'$ and $\mathcal{E}$ is denoted by $\langle , \rangle$. If $x \in E$ and $\xi \in \mathcal{E}$, then $\langle x, \xi \rangle$ is the inner product of the elements x and ξ in E. If $0 < p < q$, then we have the following continuous inclusions:

$$\mathcal{E} \subset \mathcal{E}_q \subset \mathcal{E}_p \subset E \subset \mathcal{E}'_p \subset \mathcal{E}'_q \subset \mathcal{E}'.$$

The following sequence of inclusions

$$\mathcal{E} \subset E \subset \mathcal{E}'$$

is called a Gel'fand triple.

By Minlos' theorem there exists a unique probability measure μ on the Borel subsets of $\mathcal{E}'$ such that for all $\xi \in \mathcal{E}$, the random variable $\langle \cdot, \xi \rangle$ is normally distributed with mean 0 and variance $|\xi|_0^2$. The characteristic function of μ is given by

$$\int_{\mathcal{E}'} e^{i\langle x, \xi \rangle} d\mu(x) = e^{-\frac{1}{2}|\xi|_0^2}, \qquad \forall \xi \in \mathcal{E}. \tag{1}$$

Because $\mathcal{E}$ is dense in E, we may define for any $f \in E$ a random variable $\langle \cdot, f \rangle$ on $\mathcal{E}'$. This random variable is normally distributed with mean 0 and variance $|f|_0^2$.

We call the probability space $(\mathcal{E}', \mu)$ a white noise space. We denote by (L^2) the space of all functions $\varphi : \mathcal{E}' \longrightarrow \mathbb{C}$ that are measurable and square integrable with respect to the measure μ, that means $\int_{\mathcal{E}'} |\varphi(x)|^2 d\mu(x) < \infty$. E_c denotes the complexification of E, while $E_c^{\hat{\otimes} n}$ the space of all symmetric n-tensors in the space $E_c^{\otimes n}$. Throughout this paper a hat above a tensor will denote the symmetrization of that tensor. For example if $u, v \in E_c$, then $u \hat{\otimes} v = (1/2)(u \otimes v + v \otimes u)$. For any $x \in \mathcal{E}'$ and $n \in \mathbb{N} \cup \{0\}$ we define *the Wick tensor* $: x^{\otimes n} : \ \in \mathcal{E}'^{\hat{\otimes} n}$ by the formula:

$$: x^{\otimes n} : \ = \sum_{k=0}^{[n/2]} \binom{n}{2k} (2k-1)!!(-1)^k x^{\otimes(n-2k)} \hat{\otimes} \tau^{\otimes k},$$

where τ is the so-called *trace operator*, which is the unique linear bounded operator from $\mathcal{E}_c \otimes \mathcal{E}_c$ into $\mathbb{C}$ such that

$$\langle \tau, \xi \otimes \eta \rangle = \langle \xi, \eta \rangle, \qquad \forall \xi, \eta \in \mathcal{E}_c.$$

For any $\varphi \in (L^2)$ there exists a unique sequence $\{f_n\}_{n \geq 0}$ such that $\forall n \geq 0$, $f_n \in E_c^{\hat{\otimes} n}$, and

$$\varphi(x) = \sum_{n=0}^{\infty} \langle : x^{\otimes n} :, f_n \rangle.$$

The above equality is in the (L^2)-sense and from it we can easily calculate the (L^2)-norm of the function φ. This norm is denoted by $\| \varphi \|_0$ and the following relation holds:

$$\| \varphi \|_0^2 = \sum_{n=0}^{\infty} n! \, | f_n |_0^2,$$

where $| \cdot |_0$ denotes the $E_c^{\otimes n}$-norm induced from the norm $| \cdot |_0$ on E. We define *the second quantization operator of A*, denoted by $\Gamma(A)$, as a densely defined self-adjoint operator on (L^2) in the following way: If $\varphi = \sum_{n=0}^{\infty} \langle : \cdot^{\otimes n} :, f_n \rangle$, then

$$\Gamma(A)\varphi = \sum_{n=0}^{\infty} \langle : \cdot^{\otimes n} :, A^{\otimes n} f_n \rangle.$$

The operator $\Gamma(A)$ has also a countable spectrum. Its eigenvalues satisfy the same conditions as those of the operator A except that the smallest eigenvalue of $\Gamma(A)$ is not strictly greater than 1, but equal to 1. For any $p \geq 0$, we define $(\mathcal{E}_p) = \{\varphi \in (L^2) \mid \; \| \Gamma(A)^p\varphi \|_0 < \infty\}$. If $\varphi \in (\mathcal{E}_p)$, then we define $\| \varphi \|_p := \| \Gamma(A)^p\varphi \|_0$. The space $(\mathcal{E}_p)$ equipped with the norm $\| \cdot \|_p$ is a Hilbert space. If $0 \leq p < q$, then $(\mathcal{E}_q) \subset (\mathcal{E}_p)$. We define: $(\mathcal{E}) = \bigcap_{p \geq 0}(\mathcal{E}_p)$. The space $(\mathcal{E})$ equipped with the family $\{\| \cdot \|_p\}_{p \geq 0}$ of semi-norms (in fact norms) is a nuclear space. For any $p \geq 0$, the dual of the Hilbert space $(\mathcal{E}_p)$ is isomorphic to the Hilbert space $(\mathcal{E}_{-p})$, which is the completion of the space (L^2) with respect to the norm $\| \varphi \|_{-p} := \| \Gamma(A)^{-p}\varphi \|_0$. If we denote by $(\mathcal{E})^*$ the dual of the nuclear space $(\mathcal{E})$, then $(\mathcal{E})^* = \bigcup_{p \geq 0}(\mathcal{E}_{-p})$. The elements in $(\mathcal{E})$ are called <u>test functions</u> on $\mathcal{E}'$, while the elements in $(\mathcal{E})^*$ are called <u>generalized functions</u> on $\mathcal{E}'$. The bilinear pairing between $(\mathcal{E})^*$ and $(\mathcal{E})$ is denoted by $\langle\langle \cdot, \cdot \rangle\rangle$. If $\varphi \in (L^2)$ and $\psi \in (\mathcal{E})$, then $\langle\langle \varphi, \psi \rangle\rangle = (\varphi, \overline{\psi})$, where $(\cdot, \cdot)$ is the inner product on the complex Hilbert space (L^2).

An element $\varphi \in (\mathcal{E})$ can be uniquely written in the form $\varphi = \sum_{n=0}^{\infty}\langle : \cdot^{\otimes n} :, f_n \rangle$, $f_n \in \mathcal{E}_c^{\hat{\otimes} n}$. Moreover,

$$\|\varphi\|_p^2 = \sum_{n=0}^{\infty} n!|f_n|_p^2 < \infty, \qquad \forall p \geq 0.$$

In the same way, any element $\phi \in (\mathcal{E})^*$ can be uniquely represented as $\phi = \sum_{n=0}^{\infty}\langle : \cdot^{\otimes n} :, F_n \rangle$, $F_n \in (\mathcal{E}_c')^{\hat{\otimes} n}$. If $\phi \in (\mathcal{E})^*$ has the above representation, then there exists $p \geq 0$ such that

$$\| \phi \|_{-p}^2 = \sum_{n=0}^{\infty} n! \mid F_n \mid_{-p}^2 < \infty.$$

If $\phi = \sum_{n=0}^{\infty}\langle : \cdot^{\otimes n} :, F_n \rangle \in (\mathcal{E})^*$ and $\varphi = \sum_{n=0}^{\infty}\langle : \cdot^{\otimes n} :, f_n \rangle \in (\mathcal{E})$, then the following relation holds:

$$\langle\langle \phi, \varphi \rangle\rangle = \sum_{n=0}^{\infty} n!\langle F_n, f_n \rangle.$$

(b) The Exponential Functions and The S-Transform

For any $x \in \mathcal{E}_c'$, we define the renormalized exponential function $: e^{\langle \cdot, x \rangle} :$ in the following way:

$$: e^{\langle \cdot, x \rangle} := \sum_{n=0}^{\infty} \frac{1}{n!}\langle : \cdot^{\otimes n} :, x^{\otimes n} \rangle.$$

A simple calculation shows that, for any $p \in \mathbb{R}$,

$$\| : e^{\langle \cdot, x \rangle} : \|_p = e^{\frac{|x|_p^2}{2}}. \tag{2}$$

Hence for any $x \in \mathcal{E}'_c$, we have $: e^{\langle \cdot, x \rangle} : \in (\mathcal{E})^*$. Moreover, it follows from (2) that $: e^{\langle \cdot, x \rangle} : \in (L^2)$ if and only if $x \in E_c$, and $: e^{\langle \cdot, x \rangle} : \in (\mathcal{E})$ if and only if $x \in \mathcal{E}_c$. It is also easy to check that if $x \in \mathcal{E}'_c$ and $\xi \in \mathcal{E}_c$, then

$$\langle\langle : e^{\langle \cdot, x \rangle} :, : e^{\langle \cdot, \xi \rangle} : \rangle\rangle = e^{\langle x, \xi \rangle}.$$

For any $\xi \in E_c$, we have:

$$: e^{\langle x, \xi \rangle} : = e^{\langle x, \xi \rangle - (1/2)\langle \xi, \xi \rangle}.$$

The renormalized exponential functions $\{ : e^{\langle \cdot, \xi \rangle} : \mid \xi \in \mathcal{E}_c \}$ are linearly independent and span a dense subspace of $(\mathcal{E})$.

For any $\Phi \in (\mathcal{E})^*$, the *S-transform of* Φ is the function $S(\Phi) : \mathcal{E}_c \to \mathbb{C}$ defined by

$$S(\Phi)(\xi) = \langle\langle \Phi, : e^{\langle \cdot, \xi \rangle} : \rangle\rangle, \quad \xi \in \mathcal{E}_c.$$

Since the renormalized exponential functions span a dense subspace of $(\mathcal{E})$, the S-transform is injective. That means, if $\Phi, \Psi \in (\mathcal{E})^*$ such that $S(\Phi) = S(\Psi)$, then $\Phi = \Psi$.

For $\Phi \in (L^2)$, the S-transform of Φ is also called the *Segal-Bargmann transform of* Φ.

(c) The Wick product

The Wick product of two generalized functions Φ and Ψ in $(\mathcal{E})^*$, denoted by $\Phi \diamond \Psi$, is the unique generalized function in $(\mathcal{E})^*$ such that:

$$S(\Phi \diamond \Psi) = (S\Phi)(S\Psi).$$

The mapping $(\phi, \psi) \mapsto \phi \diamond \psi$ is continuous from $(\mathcal{E}) \times (\mathcal{E})$ into $(\mathcal{E})$. It is also continuous from $(\mathcal{E})^* \times (\mathcal{E})^*$ into $(\mathcal{E})^*$. If $\varphi = \sum_{n=0}^{\infty} \langle : \cdot^{\otimes n} :, f_n \rangle \in (\mathcal{E})$ and $\psi = \sum_{n=0}^{\infty} \langle : \cdot^{\otimes n} :, g_n \rangle \in (\mathcal{E})$, then

$$\varphi \diamond \psi = \sum_{n=0}^{\infty} \langle : \cdot^{\otimes n} :, \sum_{p+q=n} f_p \hat{\otimes} g_q \rangle.$$

3 HAUSDORFF-YOUNG INEQUALITY FOR WHITE NOISE ANALYSIS

A starting point in formulating a Hausdorff-Young inequality is the following theorem established by the contribution of Kondratiev [3], Krée [4, 5, 6], and Lee [8].

Theorem 3 *The S-transform is a unitary operator from (L^2) onto the Hilbert space $\mathcal{H}L^2(E_c, m)$ consisting of all holomorphic functions $f : E_c \to \mathbb{C}$ such that $\| f \|^2 \equiv \sup\{\int_F |f(z)|^2 dm_F(z)\} < \infty$, with the supremum being taken over all finite dimensional subspaces F of E_c. If $F \equiv \mathbb{C}^n$, then $dm_F(z) = (\pi)^{-n} \exp[-|z|^2]dz$, where $dz = dxdy$, for $z = x + iy \in \mathbb{C}^n$, $x, y \in \mathbb{R}^n$.*

We introduce below two important spaces:

Definition 4 *For any strictly positive real number q, we define the space:*

$$(L^q) = \left\{ \varphi = \sum_{n=0}^{\infty} \langle : \cdot^{\otimes n} :, f_n \rangle \in (\mathcal{E})^* \mid \sum_{n=0}^{\infty} \left(\frac{q}{2}\right)^n n! |f_n|_0^2 < \infty \right\}.$$

If $\varphi = \sum_{n=0}^{\infty} \langle : \cdot^{\otimes n} :, f_n \rangle \in (L^q)$, then we define $\| \varphi \|_{(q)}^2 := \sum_{n=0}^{\infty} \left(\frac{q}{2}\right)^n n! |f_n|_0^2$.

Definition 5 *For any real number $q \geq 1$, we define the space $\mathcal{H}L^q(E_c, m)$ consisting of all holomorphic functions $f : E_c \to \mathbb{C}$ such that $\| f \|_{[q]}^q \equiv \sup\{\int_F |f(z)|^q dm_F(z)\} < \infty$, with the supremum being taken over all finite dimensional subspaces F of E_c. If $F \equiv \mathbb{C}^n$, then $dm_F(z) = (\pi)^{-n} e^{-|z|^2} dz$, where $dz = dxdy$, for $z = x + iy \in \mathbb{C}^n$, $x, y \in \mathbb{R}^n$.*

Observation 6 *For any $q_1, q_2 \in \mathbb{R}$ such that $0 < q_1 < q_2$, $(L^{q_2}) \subset (L^{q_1})$. If $\varphi \in (L^{q_2})$, then $\| \varphi \|_{(q_1)} \leq \| \varphi \|_{(q_2)}$.*

Theorem 7 *For any even natural number q, the S-transform is a bounded linear operator from (L^q) into $\mathcal{H}L^q(E_c, m)$ with norm 1.*

Proof. Let $q = 2k$, where $k \in \mathbb{N}$. Let $\varphi = \sum_{n=0}^{\infty} \langle : \cdot^{\otimes n} :, f_n \rangle \in (L^q)$. We have:

$$\| S\varphi \|_{[q]}^q = \sup \left\{ \int_F |S\varphi(z)|^q dm_F(z) \right\}$$

$$= \sup \left\{ \int_F |S(\underbrace{\varphi \diamond \varphi \diamond \cdots \diamond \varphi}_{k \text{ times}})(z)|^2 dm_F(z) \right\}.$$

By Theorem 3 we have:

$$\sup\left\{\int_F |S(\underbrace{\varphi\diamond\varphi\diamond\cdots\diamond\varphi}_{k\ \text{times}})(z)|^2 dm_F(z)\right\} = \|\underbrace{\varphi\diamond\varphi\diamond\cdots\diamond\varphi}_{k\ \text{times}}\|_0^2 .$$

Thus, we obtain:

$$\|S\varphi\|_{[q]}^q = \|\underbrace{\varphi\diamond\varphi\diamond\cdots\diamond\varphi}_{k\ \text{times}}\|_0^2$$

$$= \sum_{n=0}^{\infty} n! \left|\sum_{i_1+i_2+\cdots+i_k=n} f_{i_1}\hat\otimes f_{i_2}\hat\otimes\cdots\hat\otimes f_{i_k}\right|_0^2$$

$$\leq \sum_{n=0}^{\infty} n! \left[\sum_{i_1+i_2+\cdots+i_k=n} |f_{i_1}\hat\otimes f_{i_2}\hat\otimes\cdots\hat\otimes f_{i_k}|_0\right]_0^2 .$$

Since the symmetrization operator is a projection operator, we have:

$$|f_{i_1}\hat\otimes f_{i_2}\hat\otimes\cdots\hat\otimes f_{i_k}|_0 \leq |f_{i_1}\otimes f_{i_2}\otimes\cdots\otimes f_{i_k}|_0 .$$

Hence we get:

$$\|S\varphi\|_{[q]}^q$$

$$\leq \sum_{n=0}^{\infty} n! \left[\sum_{i_1+i_2+\cdots+i_k=n} |f_{i_1}\otimes f_{i_2}\otimes\cdots\otimes f_{i_k}|_0\right]^2$$

$$= \sum_{n=0}^{\infty} n! \left[\sum_{i_1+i_2+\cdots+i_k=n} |f_{i_1}|_0|f_{i_2}|_0\cdots|f_{i_k}|_0\right]^2$$

$$= \sum_{n=0}^{\infty} \left[\sum_{i_1+i_2+\cdots+i_k=n} \sqrt{n!}\,|f_{i_1}|_0|f_{i_2}|_0\cdots|f_{i_k}|_0\right]^2$$

$$= \sum_{n=0}^{\infty} \left[\sum_{i_1+i_2+\cdots+i_k=n} \sqrt{\frac{n!}{i_1!i_2!\cdots i_k!}}\sqrt{i_1!}|f_{i_1}|_0\sqrt{i_2!}|f_{i_2}|_0\cdots\sqrt{i_k!}|f_{i_k}|_0\right]^2 .$$

Applying the Cauchy-Bunyakovskii-Schwarz inequality we get:

$$\sum_{n=0}^{\infty} \left[\sum_{i_1+i_2+\cdots+i_k=n} \sqrt{\frac{n!}{i_1!i_2!\cdots i_k!}}\sqrt{i_1!}|f_{i_1}|_0\sqrt{i_2!}|f_{i_2}|_0\cdots\sqrt{i_k!}|f_{i_k}|_0\right]^2$$

$$\leq \sum_{n=0}^{\infty}\sum_{i_1+i_2+\cdots+i_k=n} \frac{n!}{i_1!i_2!\cdots i_k!} \sum_{i_1+i_2+\cdots+i_k=n} i_1!|f_{i_1}|_0^2 i_2!|f_{i_2}|_0^2\cdots i_k!|f_{i_k}|_0^2 .$$

122

Hence we obtain:

$$\| S\varphi \|_{[q]}^{q}$$

$$\leq \sum_{n=0}^{\infty} \sum_{i_1+i_2+\cdots+i_k=n} \frac{n!}{i_1!i_2!\cdots i_k!} \sum_{i_1+i_2+\cdots+i_k=n} i_1!|f_{i_1}|_0^2 i_2!|f_{i_2}|_0^2 \cdots i_k!|f_{i_k}|_0^2$$

$$= \sum_{n=0}^{\infty} \underbrace{(1+1+\cdots+1)}_{k \text{ times}}^{n} \sum_{i_1+i_2+\cdots+i_k=n} i_1!|f_{i_1}|_0^2 \ i_2!|f_{i_2}|_0^2 \ \cdots \ i_k!|f_{i_k}|_0^2$$

$$= \sum_{n=0}^{\infty} k^n \sum_{i_1+i_2+\cdots+i_k=n} i_1!|f_{i_1}|_0^2 \ i_2!|f_{i_2}|_0^2 \ \cdots \ i_k!|f_{i_k}|_0^2$$

$$= \sum_{n=0}^{\infty} \sum_{i_1+i_2+\cdots+i_k=n} k^{i_1}i_1!|f_{i_1}|_0^2 \ k^{i_2}i_2!|f_{i_2}|_0^2 \ \cdots \ k^{i_k}i_k!|f_{i_k}|_0^2$$

$$= \prod_{j=1}^{k} \sum_{i_j=0}^{\infty} k^{i_j}i_j!|f_{i_j}|_0^2$$

$$= \left[\sum_{i=0}^{\infty} \left(\frac{q}{2}\right)^{i} i!|f_i|_0^2 \right]^{k}$$

$$= \| \varphi \|_{(q)}^{q} .$$

Thus we obtain:

$$\| S\varphi \|_{[q]} \leq \| \varphi \|_{(q)} . \tag{3}$$

This proves that the linear operator $S : (L^q) \to \mathcal{H}L^q(E_c, m)$ is continuous and its operatorial norm $\| S \|$ satisfies the inequality $\| S \| \leq 1$. To see that $\| S \| = 1$, let's consider the function $\varphi = e^{\langle \cdot, \eta \rangle}$, where $\eta \in E$, $|\eta|_0 = 1$. Let F be the one dimensional complex vector subspace of E_c spanned by η. We have:

$$\| \varphi \|_{(q)}^{2} = \sum_{n=0}^{\infty} \left(\frac{q}{2}\right)^{n} n! \left| \frac{\eta^{\otimes n}}{n!} \right|_0^2$$

$$= \sum_{n=0}^{\infty} \left(\frac{q}{2}\right)^{n} \frac{|\eta|_0^{2n}}{n!}$$

$$= e^{\frac{q|\eta|_0^2}{2}}$$

$$= e^{\frac{q}{2}} .$$

Hence $\| \varphi \|_{(q)} = e^{\frac{q}{4}}$. We have:

$$\| S \| \cdot \| \varphi \|_{(q)} \geq \| S\varphi \|_{[q]}$$

$$\geq \left[\frac{1}{\pi} \int_F |S\varphi(\xi)|^q d\mu_F(\xi) \right]^{1/q}$$

$$= \left[\frac{1}{\pi} \int_F |e^{\langle \eta, \xi \rangle}|^q d\mu_F(\xi) \right]^{1/q}$$

$$= \left[\frac{1}{\pi} \int_{\mathbb{C}} |e^{\langle \eta, z\eta \rangle}|^q e^{-|z\eta|_0^2} dz \right]^{1/q}$$

$$= \left[\frac{1}{\pi} \int_{\mathbb{C}} |e^z|^q e^{-|z|^2} dz \right]^{1/q}$$

$$= \left[\frac{1}{\pi} \int_{\mathbb{R}} \int_{\mathbb{R}} |e^{x+iy}|^q e^{-x^2-y^2} dx dy \right]^{1/q}$$

$$= \left[\frac{1}{\pi} \int_{\mathbb{R}} \int_{\mathbb{R}} e^{qx} e^{-x^2-y^2} dx dy \right]^{1/q}$$

$$= \left[\frac{1}{\sqrt{\pi}} \int_{\mathbb{R}} e^{\frac{q^2}{4}} e^{-\frac{q^2}{4}+qx-x^2} dx \, \frac{1}{\sqrt{\pi}} \int_{\mathbb{R}} e^{-y^2} dy \right]^{1/q}$$

$$= e^{\frac{q}{4}} \left[\frac{1}{\sqrt{\pi}} \int_{\mathbb{R}} e^{-\left(x-\frac{q}{2}\right)^2} dx \right]^{1/q}$$

$$= e^{\frac{q}{4}}$$

$$= \| \varphi \|_q \, .$$

Hence $\| S \| \geq 1$ and so $\| S \| = 1$. Therefore we have proved the theorem for any $q \in \{2, 4, 6, \cdots\}$. $\qquad\square$

If $B : E \to E$ is a bounded linear operator, then we may define *the second quantization operator of B*, denoted by $\Gamma(B)$, as a densely-defined operator on (L^2), in the following way: if $\varphi = \sum_{n=0}^{\infty} \langle : \cdot^{\otimes n} :, f_n \rangle$, then

$$\Gamma(B)\varphi = \sum_{n=0}^{\infty} \langle : \cdot^{\otimes n} :, B^{\otimes n} f_n \rangle.$$

Observation 8 *If I denotes the identity operator on E, then for any $q \geq 2$ and $\varphi \in (L^q)$, we have*

$$\| \varphi \|_{(q)} = \left\| \Gamma\left(\sqrt{\frac{q}{2}} I\right) \varphi \right\|_0 . \tag{4}$$

124

Thus the last theorem can be rephrased in the following way: for any even natural number q and for any $\varphi \in (L^q)$,

$$\| S\varphi \|_{[q]} \leq \left\| \Gamma\left(\sqrt{\tfrac{q}{2}}I\right)\varphi \right\|_0 . \tag{5}$$

The following theorem is an analog of the classical Young inequality.

Theorem 9 *Let p, q, and r be strictly positive numbers such that $\frac{1}{p} + \frac{1}{q} = \frac{1}{r}$. If $\varphi \in (L^p)$ and $\psi \in (L^q)$, then $\varphi \diamond \psi \in (L^r)$ and the following inequality holds:*

$$\| \varphi \diamond \psi \|_{(r)} \leq \| \varphi \|_{(p)} \| \psi \|_{(q)} . \tag{6}$$

Proof. Let $\varphi = \sum_{n=0}^{\infty} \langle : \cdot^{\otimes n} :, f_n \rangle \in (L^p)$ and $\psi = \sum_{n=0}^{\infty} \langle : \cdot^{\otimes n} :, g_n \rangle \in (L^q)$. Then $\varphi \diamond \psi = \sum_{n=0}^{\infty} \langle : \cdot^{\otimes n} :, h_n \rangle$, where $h_n = \sum_{u+v=n} f_u \hat{\otimes} g_v$. We have:

$$\| \varphi \diamond \psi \|_{(r)}^2$$

$$= \sum_{n=0}^{\infty} \left(\frac{r}{2}\right)^n n! |h_n|_0^2$$

$$= \sum_{n=0}^{\infty} \left(\frac{r}{2}\right)^n n! \left| \sum_{u+v=n} f_u \hat{\otimes} g_v \right|_0^2$$

$$\leq \sum_{n=0}^{\infty} \left(\frac{r}{2}\right)^n n! \left[\sum_{u+v=n} |f_u \hat{\otimes} g_v|_0 \right]^2$$

$$\leq \sum_{n=0}^{\infty} \left(\frac{r}{2}\right)^n n! \left[\sum_{u+v=n} |f_u \otimes g_v|_0 \right]^2$$

$$= \sum_{n=0}^{\infty} \left(\frac{r}{2}\right)^n n! \left[\sum_{u+v=n} |f_u|_0 |g_v|_0 \right]^2$$

$$= \sum_{n=0}^{\infty} \left(\frac{r}{2}\right)^n \left[\sum_{u+v=n} \sqrt{\frac{n!}{u!v!}} \left(\frac{2}{p}\right)^u \left(\frac{2}{q}\right)^v \sqrt{\left(\frac{p}{2}\right)^u u!} |f_u|_0 \sqrt{\left(\frac{q}{2}\right)^v v!} |g_v|_0 \right]^2 .$$

Applying the Cauchy-Bunyakovskii-Schwarz inequality we obtain:

$$\sum_{n=0}^{\infty} \left(\frac{r}{2}\right)^n \left[\sum_{u+v=n} \sqrt{\frac{n!}{u!v!}} \left(\frac{2}{p}\right)^u \left(\frac{2}{q}\right)^v \sqrt{\left(\frac{p}{2}\right)^u u!} |f_u|_0 \sqrt{\left(\frac{q}{2}\right)^v v!} |g_v|_0 \right]^2$$

$$\leq \sum_{n=0}^{\infty} \left(\frac{r}{2}\right)^n \sum_{u+v=n} \frac{n!}{u!v!} \left(\frac{2}{p}\right)^u \left(\frac{2}{q}\right)^v \sum_{u+v=n} \left(\frac{p}{2}\right)^u u! |f_u|_0^2 \left(\frac{q}{2}\right)^v v! |g_v|_0^2.$$

Thus, we have:

$$\| \varphi \diamond \psi \|_{(r)}^2$$

$$\leq \sum_{n=0}^{\infty} \left(\frac{r}{2}\right)^n \sum_{u+v=n} \frac{n!}{u!v!} \left(\frac{2}{p}\right)^u \left(\frac{2}{q}\right)^v \sum_{u+v=n} \left(\frac{p}{2}\right)^u u! |f_u|_0^2 \left(\frac{q}{2}\right)^v v! |g_v|_0^2$$

$$= \sum_{n=0}^{\infty} \left(\frac{r}{2}\right)^n \left(\frac{2}{p}+\frac{2}{q}\right)^n \sum_{u+v=n} \left(\frac{p}{2}\right)^u u! |f_u|_0^2 \left(\frac{q}{2}\right)^v v! |g_v|_0^2$$

$$= \sum_{n=0}^{\infty} \left(\frac{r}{2}\right)^n \left(\frac{2}{r}\right)^n \sum_{u+v=n} \left(\frac{p}{2}\right)^u u! |f_u|_0^2 \left(\frac{q}{2}\right)^v v! |g_v|_0^2$$

$$= \sum_{n=0}^{\infty} \sum_{u+v=n} \left(\frac{p}{2}\right)^u u! |f_u|_0^2 \left(\frac{q}{2}\right)^v v! |g_v|_0^2$$

$$= \sum_{u=0}^{\infty} \left(\frac{p}{2}\right)^u u! |f_u|_0^2 \sum_{v=0}^{\infty} \left(\frac{q}{2}\right)^v v! |g_v|_0^2$$

$$= \| \varphi \|_{(p)}^2 \| \psi \|_{(q)}^2 .$$

$\square$

References

1. Bargmann, V.: On a Hilbert space of analytic functions and an associated integral transform, Part I; *Communications of Pure and Applied Mathematics* **24** (1961) 187-214.

2. Gross, L. and Malliavin, P.: Hall's transform and the Segal-Bargmann map; in "Itô's Stochastic Calculus and Probability Theory", N. Ikeda, S. Watanabe, M. Fukushima, and H. Kunita (eds.), 73-116, Springer-Verlag, 1996.

3. Kondratiev, Yu. G.: Nuclear spaces of entire functions in problems of infinite-dimensional analysis; *Soviet Math. Dokl.* **22** (1980) 588-592.

4. Krée, P.: Solutions faibles d'équations aux d'érivées functionnelles, I; *Lecture Notes in Math.* **410** (1974) 142-181, Springer-Verlag.

5. Krée, P.: Solutions faibles d'équations aux d'érivées functionnelles, II; *Lecture Notes in Math.* **474** (1975) 16-47, Springer-Verlag.

6. Krée, P.: Calcul d'intégrales et de dérivées en dimension infinie; *J. Funct. Anal.* **31** (1979) 150-186.

7. Kuo, H. H.: *White Noise Distribution Theory*, Probability and Stochastic Series, CRC Press, Inc. 1996.

8. Lee, Y.-J.: A characterization of generalized functions on infinite-dimensional spaces and Bargmann-Segal analytic functions; in "Gaussian random Fields", K. Itô and T. Hida (eds.) (1991) 272-284, World Scientific.

9. Obata, N.: *White Noise Calculus on Fock Space*, Lecture Notes in Math. **1577**, Springer-Verlag (1994).

10. Reed, M. and Simon, B.: *Methods of Modern Mathematical Physics II: Fourier Analysis, Self-adjointness*, Academic Press, 1975.

11. Segal, I. E.: Tensor algebras over Hilbert spaces, *Trans. Amer. Math. Soc.* **81** (1956) 106-134.

12. Segal, I. E.: Mathematical characterization of the physical vacuum for a linear Bose-Einstein field, *Illinois J. Math.* **6** (1962) 500-523.

13. Segal, I. E.: The complex wave representation of the free Boson field; in "Topics in functional analysis: essays dedicated to M.G. Krein on the occasion of his 70th birthday", Advances in Mathematics: Supplementary Studies (I. Gohberg and M. Kac, eds.), vol.3, Academic Press, New York, 1978, pp. 321-344.

Quantum Information IV (pp. 127–145)
Eds. T. Hida and K. Saitô
© 2002 World Scientific Publishing Co.

THE CLARK FORMULA OF GENERALIZED WIENER FUNCTIONALS

YUH-JIA LEE*

Department of Applied Mathematics
National University of Kaohsiung
Kaohsiung, TAIWAN 811

HSIN-HUNG SHIH†

Center of General Education
Kung Shan University of Technology
Tainan, TAIWAN 710

The Clark formula is reformulated on the classical Wiener space for a large class of generalized Wiener functional via the idea of white noise analysis. The underlying spcae is taken to be the Gel'fand triple $E \subset C' \subset E^*$ associated with the space C' of Cameron-Martin functions. Then we adapt the space $\mathcal{E}_0$ of analytic function of exponential growth as the space of test functionals in the construction of the space $\mathcal{E}_0^*$ of generalized Wiener functionals. It is shown that the S-transform SF of a generalized Wiener functionals F satifies the following formula

$$SF(\eta) = \langle\langle F, 1 \rangle\rangle_\infty + \int_0^1 \frac{d}{dt} SF(P_t(\eta))dt,$$

which follows easily from the fundamental theorem of calculus, where $\eta \in E_c$, the complexification of E, and $P_t(\eta)(s) = \eta(\min\{s, t\})$. The the Clark formula follows immediately.

1 Introduction

The representation of functionals of Brownian motion by stochastic integral with respect to Brownian motion, known as the Clark formula, was first studied in [1] and later in [20, 5]. It was generalized to weakly differentiable functionals on the classical Wiener space C by Ocone [20] via Malliavin calculus. Later, it was generalized to Malliavin distributions in [6, 22]. The first reformulation of the Clark formula using Hida calculus (or white noise analysis) has been done in [2].

In this paper we are devoted to the reformulation of the Clark formula on the classical Wiener space for a class of generalized Wiener functional via

---

*,†RESEARCH SUPPORTED BY THE NATIONAL SCIENCE COUNCIL OF TAIWAN

the idea of white noise analysis. The underlying spcae is taken to be the Gel'fand triple $E \subset C' \subset E^*$ with a chain of Hilbert spaces E_p with the norm $|x|_p = |K^{-p/2}x|_0$, where C' is the Cameron-Martin space,

$$Kx(t) = \int_0^1 \min\{s, t\}\, x(s)\, ds, \quad x \in L^2[0, 1],$$

E_p is the domain of $K^{-p/2}$ for any $p \geq 0$, and $E = \cap_{p \geq 0} E_p$ is the projective limit of E_p's. The dual space of E_p is denoted by E_{-p}.

To construct the generalized Wiener functional, for $p \geq 1$, let $\mathcal{E}_p$ be the space of functionals f defined on E_{-p} such that

(1) f has a analytic extension $\widetilde{f}$ to the complexification $E_{-p,c}$ of E_{-p},

(2) There exist constants c, c' such that $|\widetilde{f}(z)| \leq c \exp\{c'|z|_{-p}\}$ for $z \in E_{-p,c}$.

Denote by $\mathcal{E}_0$ the corresponding space defined on C similarly as above .

Set $\mathcal{E}_\infty = \cap_{p=1}^\infty \mathcal{E}_p$. $\mathcal{E}_\infty$ is the collection of analytic version of members of the Yan-Meyer space and we have the follwing relations

$$\mathcal{E}_\infty \subset \mathcal{E}_p \subset \mathcal{E}_0 \subset L^2[C, \mu] \subset \mathcal{E}_0^* \subset \mathcal{E}_p^* \subset \mathcal{E}_\infty^*,$$

where μ is the Wiener measure on C and $\mathcal{E}_\infty{}^*$ denote the dual space of $\mathcal{E}_\infty$. Members of $\mathcal{E}_p^*$ and $\mathcal{E}_\infty^*$ are called generalized Wiener functionals.

In this paper we show that the S-transform SF of a generalized Wiener functionals $F \in \mathcal{E}_0^*$ satisfies the following formula

$$SF(\eta) = \langle\!\langle F,\, 1 \rangle\!\rangle_\infty + \int_0^1 \frac{d}{dt}\, SF(P_t(\eta))\, dt,$$

which follows easily from the fundamental theorem of calculus, where $\eta \in E_c$, the complexification of E and $P_t(\eta)(s) = \eta(\min\{s, t\})$. The Clark formula of generalized Wiener functionals in $\mathcal{E}_0^*$ follows immediately. In particular, for $\varphi \in \mathcal{E}_1$, we obtain

$$\varphi(x) = \ \mathrm{E}[\varphi] + \int_0^1 \mathrm{E}[\, D\varphi(x)\, 1_{[t,1]} \,|\, \mathcal{F}_t\,]\, dB(t;\, x) \quad \mu\text{-almost all } x \text{ on } C,$$

where the integral is the Wiener-Itô stochastic integral.

The following notations will be used frequently in this paper.

Notations.

1. For a real linear space V, V_c denotes the complexification of V.

2. For each continuous function x on $[0,1]$, $\dot{x}(t)$, $\ddot{x}(t)$, $x^{(p)}(t)$ are respectively the first, second, and p-th order derivatives of x with respect to $t \in [0,1]$, if they exist.

3. For a n-linear operator T on a complex normed space X, $Tz^n :=$ $T(z,\dots,z)$ (n copies) for each $z \in X$.

4. For any Hilbert space H, $\|\cdot\|_{HS^n(H)}$ denotes the n-linear Hilbert-Schmidt operator norm and $\|\cdot\|_{\mathcal{L}^n(H)}$ the n-linear operator norm over H.

2   A Gel'fand triple associated with the classical Wiener space

Let $\mathcal{C}$ be the collection of all real-valued continuous functions x on $[0,1]$ vanishing at zero and the subclass $\mathcal{C}'$ of $\mathcal{C}$ consist of all absolutely continuous functions x with $\dot{x} \in L^2[0,1]$. Then $\mathcal{C}$ is a Banach space with the sup-norm $|\cdot|_\infty$ and $\mathcal{C}'$ is a Hilbert space with norm $|\cdot|_0 = \sqrt{\langle\cdot,\cdot\rangle_0}$ and the inner product $\langle\cdot,\cdot\rangle_0$ defined by $\langle x, y\rangle_0 = \int_0^1 \dot{x}(t)\,\dot{y}(t)\,dt$. The space $\mathcal{C}'$ is usually called the Cameron-Martin space and it is well known that $(\mathcal{C}',\mathcal{C})$ forms an abstract Wiener space (AWS, for abbreviation). The dual space $\mathcal{C}^*$ of $\mathcal{C}$, may be identified as a dense subspace of $\mathcal{C}'$ given as follows

Fact 2.1. [23,17]
$\mathcal{C}^* = \{\, x \in \mathcal{C}' : \dot{x} \text{ is right continuous and of bounded variation with } \dot{x}(1) = 0 \,\}$.
By this identification, whenever $x \in \mathcal{C}$ and $y \in \mathcal{C}^*$,

$$(x,\,y) = -\int_0^1 x(t)\,d\dot{y}(t),$$

where $(\cdot,\cdot)$ is the $\mathcal{C}$-$\mathcal{C}^*$ pairing. In particular, if $x \in \mathcal{C}'$, $(x,\,y) = \langle x,\,y\rangle_0$.

Let K be a bounded linear operator on $L^2[0,1]$ defined by

$$Kx(t) = \int_0^1 (s \wedge t)\,x(s)\,ds, \quad x \in L^2[0,1],$$

where $s \wedge t$ denotes the minimum of s and t. Then K is a positive self-adjoint compact operator having a complete orthonormal basis (CONS, for abbreviation) consisting of functions $\{e_n(t) = \sqrt{2}\,\sin(n-1/2)\pi t : n = 1,2,\dots\}$ with the corresponding eigenvalues given by $\{\,1/((n-1/2)\pi)^2 : n = 1,2,\dots\}$. For $n \in \mathbb{N}$, let

$$f_n(t) = \sqrt{K}\,e_n(t), \quad t \in [0,1]. \tag{2.1}$$

Observe that, for each $n \in \mathbb{N}$ and for $t \in [0,1]$, $e_n(1-t) = (-1)^{n+1}\dot{f}_n(t)$, then, by Fact 2.1, $\{f_n : n \in \mathbb{N}\} \subset \mathcal{C}^*$. Moreover, $\{f_n : n \in \mathbb{N}\}$ is also a CONS of $L^2[0,1]$. Thus, $\{f_n : n \in \mathbb{N}\}$ is a CONS of C'. Let A be the inverse operator of $\sqrt{K}$ on C'. Then A is self-adjoint densely defined on C' and $A f_n = \lambda_n f_n$ for each $n \in \mathbb{N}$, where $\lambda_n = (n - 1/2)\pi$

For any $p \geq 0$, let E_p be the domain of A^p. Then E_p is a real Hilbert space with the norm $|x|_p = |A^p x|_0$ ($C' = E_0$) and $\{(1/\lambda_n^p) f_n : n \in \mathbb{N}\}$ forms a CONS of E_p. The increasing family $\{|\cdot|_p : p \geq 0\}$ of norms are compatible and comparable; and the embedding from $E_{p+\alpha}$ into E_p is of Hilbert-Schmidt type whenever $\alpha > 1/2$. Next, let E_{-p} be the completion of C' with respect to the norm $|x|_{-p} = |A^{-p} x|_0$. Then E_{-p} is a Hilbert spaces with a CONS $\{\lambda_n^p f_n : n \in \mathbb{N}\}$. Identify $x \in E_p^*$ with $\sum_{n=1}^{\infty} (x, f_n) f_n \in E_{-p}$, where $(\cdot, \cdot)$ is the E_p^*-E_p pairing. E_{-p} becomes the dual space of E_p. Set $E = \cap_{p \geq 0} E_p$ as the projective limit of E_p's. Then E is a nuclear space with the dual $E^* = \cup_{p>0} E_{-p}$ and $E \subset C' \subset E^*$ forms a Gel'fand triple .

Observe that, for $x \in C'$ and $p \geq 1$,

$$|x|_{-p}^2 = \sum_{n=1}^{\infty} \lambda_n^{-2p} \langle x, f_n \rangle_0^2 = \sum_{n=1}^{\infty} \lambda_n^{-2(p-1)} \left(\int_0^1 x(t)\, e_n(t)\, dt \right)^2$$

$$\leq \int_0^1 x(t)^2\, dt \leq |x|_\infty^2.$$

It follows that $C \subset E_{-p}$ for all $p \geq 1$. Furthermore, by the denseness of C' in C and in E_{-p}, we have the following chain of continuous inclusion:

$$E \subset E_p \subset E_q \subset E_1 = L_2^* \subset C^* \subset C' \subset C \subset L_2 = E_{-1} \subset E_{-q} \subset E_{-p} \subset E^*,$$

where $p \geq q \geq 1$ and L_2 denotes the space $L^2[0,1]$. For notational convenience, we will use the notation $(\cdot, \cdot)$ to stand for all the dual pairings of E^*-E, E_{-p}-E_p ($p \geq 1$), and C-C^*.

Let i_0 and i_{-p} be respectively the embeddings from C' into C and E_{-p} with $p \geq 1$. Then (C', E_{-p}, i_{-p}) is a AWS. Let μ and μ_{-p}, $p \geq 1$, be the abstract Wiener measures of C and E_{-p} respectively. Then the measurable support of μ_{-p} is contained in C and, for any integrable complex-valued function φ on (E_{-p}, μ_{-p}), the restriction of φ on C is $\mathcal{B}(C)$-measurable and

$$\int_{E_{-p}} \varphi(x)\, \mu_{-p}(dx) = \int_C \varphi(x)\, \mu(dx).$$

Proposition 2.2. *For $p \in \mathbb{N}$, E_p is the class consisting of all functions $x \in C'$ with the property: (i) $\dot{x}, \ddot{x}, \ldots, x^{(p)}$ are absolutely continuous with $x^{(p+1)} \in L^2[0,1]$ and (ii) $x^{(2k)}(0) = x^{(2k+1)}(1) = 0$ for $k = 0, 1, \ldots, [p/2]$. Moreover,*

$$(x, y) = -\int_0^1 x(t)\, d\dot{y}(t) \quad \text{for } x \in L_2 \text{ and } y \in E_p. \tag{2.2}$$

Proof. Denote by U the class of functions satisfying the conditions (i) and (ii). We shall show that $U = E_p$.

Let y be in E_p. Then $y = \sum_{n=1}^{\infty} \langle y, f_n \rangle_0 f_n$ in $E_p(\subset C')$, where f_n's are defined as in (2.1). Since $\ddot{e}_n = \lambda_n^2 e_n$ for each $n \in \mathbb{N}$,

$$\dot{y}(t) = \sum_{n=1}^{\infty} (1/\lambda_n) \langle y, f_n \rangle_0 \, \dot{e}_n(t) \quad (\text{in } L_2)$$

$$= \sum_{n=1}^{\infty} \lambda_n \langle y, f_n \rangle_0 \int_t^1 e_n(s) \, ds \quad (\text{in } L_2), \quad t \in [0, 1].$$

Let $y_1 = \sum_{n=1}^{\infty} \lambda_n \langle y, f_n \rangle_0 \, e_n$. Then $y_1 \in L_2$ and $\dot{y}(t) = \int_t^1 y_1(s) \, ds$ for all $t \in [0, 1]$. Thus $\dot{y}$ is absolutely continuous with $\dot{y}(1) = 0$ and $\ddot{y} = y_1$. As $p \geq 2$, let $y_2 = \sum_{n=1}^{\infty} \lambda_n \langle y, f_n \rangle_0 \, \dot{e}_n$. Then this series converges to y_2 in L_2 and

$$\ddot{y}(t) = y_1(t) = \int_0^t y_2(s) \, ds, \quad t \in [0, 1].$$

It implies that $\ddot{y}$ is absolutely continuous with $\ddot{y}(0) = 0$ and $y^{(3)} = y_2 \in L_2$. Continuing in this way, it is easy to see that $E_p \subset U$.

To show that $U \subset E_p$, let $y \in U$ be arbitrarily given. Then $y^{(p+1)} \in L^2[0, 1]$ and hence

$$\sum_{n=1}^{\infty} \left| \int_0^1 y^{(p+1)}(t) \, \sqrt{2} \, \cos\left(\frac{1 - (-1)^p}{4} \pi - (n - 1/2)\pi t \right) dt \right|^2 < +\infty.$$

Applying integration by parts formula, we see that $\sum_{n=1}^{\infty} \lambda_n^{2p} \langle y, f_n \rangle_0^2 < +\infty$. That is, $y \in E_p$. This proves that $U = E_p$. Finally, from Fact 2.1 it follows that for $x \in C$ and $y \in E_p$,

$$(x, y) = \int_0^1 x(t) \, d\dot{y}(t) = -\int_0^1 x(t) \, \ddot{y}(t) \, dt.$$

Extending x to the whole space L_2 by continuity, we obtain the formula (2.2). $\qquad \square$

Corollary 2.3. *The nuclear space E consists of all real-valued almost everywhere differentiable functions x defined on $[0, 1]$ suchn that $x^{(2k)}(0) = x^{(2k+1)}(1) = 0$ for each $k \in \mathbb{N} \cup \{0\}$.*

It follows from the definition of the $|\cdot|_p$-norm that we have

Corollary 2.4. *(i) For $p \in \mathbb{N}$ and $x \in E_p$, $|x|_p^2 = \int_0^1 |x^{(p+1)}(t)|^2\, dt$.*

(ii) For $x \in L_2$ and $p \in \mathbb{N}$,

$$|x|_{-p-1}^2 = \int_0^1 \left| \int_{a_p}^{b_p} \cdots \int_{t_3}^1 \int_0^{t_2} \int_{t_1}^1 x(s)\, ds\, dt_1 dt_2 \cdots dt_{p-1} \right|^2 dt_p,$$

where $a_p = t_p$ and $b_p = 1$ if p is odd; $a_p = 0$ and $b_p = t_p$ if p is even.

3 The spaces of test and generalized functions

If V is a real normed space with the $|\cdot|_V$-norm, then the $|\cdot|_{V_c}$-norm of V_c is given by $|x + iy|_{V_c} = \sup\{\|e^{i\theta}(x + iy)\|_{V_c} : \theta \in [0, 2\pi]\}$ for $x, y \in V$, where $\|x + iy\|_{V_c}^2 = |x|_V^2 + |y|_V^2$. In general, $\|\cdot\|_{V_c}$ is a quasi-norm and for any $x, y \in V$, we have the inequality:

$$\|x + iy\|_{V_c} \leq |x + iy|_{V_c} \leq \sqrt{2}\,\|x + iy\|_{V_c}.$$

Note that when V is a Hilbert space, the quasi-norm $\|\cdot\|_{V_c}$ coincides with the $|\cdot|_{V_c}$-norm.

In this section, we construct test functionals on the AWS $(\mathcal{C}', \mathcal{C})$ and the AWS $(\mathcal{C}', E_{-p})$ for $p \geq 1$. For notational simplicity, we use the symbols $|\cdot|_\infty$ and $|\cdot|_{-p}$ to stand also for $|\cdot|_{\mathcal{C}_c}$ and $|\cdot|_{E_{-p,c}}$, respectively. For a fixed Banach space B which is either $\mathcal{C}$ or $\mathcal{E}_{-p}$, let $\mathcal{E}(B)$ be the class of those functions φ defined on $\mathcal{C}$ so that φ has an analytic extension $\widetilde{\varphi}(z)$ to B_c and satisfies the exponential growth condition: $|\varphi(z)| \leq c \exp\{c'|z|_{B_c}\}$ for some constants $c, c' > 0$. It is clear that $\mathcal{E}(B) \subset L^2(\mathcal{C}, \mu)$ by the Fernique theorem (see [7]). For $m \in \mathbb{N}$ and $\varphi \in \mathcal{E}(B)$, define

$$\|\varphi\|_{\mathcal{E}_m(B)} = \sup\{|\widetilde{\varphi}(z)|\, e^{-m\,|z|_{B_c}} : z \in B_c\}.$$

Let $\mathcal{E}_m(B) = \{\varphi \in \mathcal{E}(B) : \|\varphi\|_{\mathcal{E}_m(B)} < +\infty\}$. Then $\{(\mathcal{E}_m(B), \|\cdot\|_{\mathcal{E}_m(B)})\}$ is an increasing sequence of Banach spaces and $\mathcal{E}(B) = \cup_{m \in \mathbb{N}} \mathcal{E}_m(B)$. Endow $\mathcal{E}(B)$ with the inductive limit topology induced by the family $\{\mathcal{E}_m(B)\}$. Then $\mathcal{E}(B)$ becomes a locally convex topological algebra. For notational simplicity, let $\mathcal{E}_{m,p} = \mathcal{E}_m(E_p)$, $\mathcal{E}_{m,0} = \mathcal{E}_m(\mathcal{C})$, $\mathcal{E}_p = \mathcal{E}(E_{-p})$, and $\mathcal{E}_0 = \mathcal{E}(\mathcal{C})$. Then we have the following chain of continuous inclusions:

$$\mathcal{E}_\infty = \cap_{p \geq 1} \mathcal{E}_p \subset \mathcal{E}_p \subset \mathcal{E}_q \subset \mathcal{E}_0 \subset L^2[\mathcal{C}, \mu] \quad \text{as } p \geq q \geq 1,$$

where $\mathcal{E}_\infty$ is topologized as the projective limit of $\{\mathcal{E}_p\}$. Thus a sequence $\{\varphi_n\}$ in $\mathcal{E}_\infty$ converges to φ in $\mathcal{E}_\infty$ iff for any $p \geq 1$, φ_n conveges to φ in $\mathcal{E}_p$ as $n \to \infty$. $\mathcal{E}_\infty$ will serve as the space of test functions for the generalized functions in our investigation.

For $f \in L^2[\mathcal{C}, \mu]$, the Wiener-Itô decomposition theorem (see [8]) assures that f can be decomposed into an orthogonal direct sum of multiple Wiener integrals $I_n(f)$ of order n, $n \in \mathbb{N}$. Let $\mu * f$ be the convolution of f and μ defined on $\mathcal{C}'$, i.e., $\mu * f(h) = \int_{\mathcal{C}} f(h + y)\,\mu(dy)$ for $h \in \mathcal{C}'$. Then $\mu * f$ is infinitely Fréchet-differentiable in the directions of $\mathcal{C}'$ and $I_n(f)$ can be represented by

$$I_n(f) = L^2[\mathcal{C}, \mu]\text{-}\lim_{k \to \infty} \frac{1}{n!} \int_{\mathcal{C}} D^n \mu * f(0)(P_k(\cdot) + i\,P_k(y))^n \,\mu(dy)$$

(see [10, 12]), where D^n denotes the n-th Fréchet derivative in the directions of $\mathcal{C}'$ and $P_k(z) = \sum_{j=1}^{k} (z, f_j)\, f_j$ for $z \in E_c^*$. Moreover,

$$\int_{\mathcal{C}} |f(x)|^2\, \mu(dx) = \sum_{n=0}^{\infty} \frac{1}{n!} \|D^n \mu * f(0)\|^2_{HS^n(\mathcal{C}')}.$$

Next, we briefly describe the space (E) of test functionals, which was introduced by Meyer and Yan (see [16, 19]). For $m \in \mathbb{N}$ and $p \geq 1$, define the $\|\cdot\|_{m,p}$-norm on $L^2[\mathcal{C}, \mu]$ by

$$\|f\|^2_{m,p} = \sum_{n=0}^{\infty} \frac{1}{m^{2n}} \|D^n \mu * f(0)\|^2_{HS^n(E_{-p})}. \tag{3.1}$$

Denote by $(E)_{m,p}$ the class of functions f in $L^2[\mathcal{C}, \mu]$ so that $\|f\|_{m,p} < +\infty$. Then $\{(E)_{m,p} : m \in \mathbb{N}\}$ is an increasing sequence of Hilbert spaces of with the inner products induced by the $\|\cdot\|_{m,p}$-norms. Let $(E)_p$ be the inductive limit of $\{(E)_{m,p}\}$ and $(E) = \cap_{p \geq 1} (E)_p$, equipped with the projective limit topology. Then the following chain of continuous inclusions hold:

$$(E) \subset (E)_p \subset (E)_q \subset L^2[\mathcal{C}, \mu] \quad \text{whenever } p \geq q \geq 1.$$

Obviously, each member f of (E) satisfies

$$\sum_{n=0}^{\infty} \frac{1}{n!} \|D^n \mu * f(0)\|^2_{HS^n(E_{-p})} < +\infty \quad \text{for all } p \geq 1.$$

According to the work by Lee [13, 15], there exists a unique analytic function $\tilde{f}$ defined on E_c^* such that $\tilde{f} = f$ μ-almost everywhere on $\mathcal{C}_c$ and

$$\tilde{f}(z) = \sum_{n=0}^{\infty} \frac{1}{n!} \int_{\mathcal{C}} D^n \mu * \tilde{f}(z + i\,y)^n \,\mu(dy), \quad z \in E_c^*,$$

where the series converges absolutely and uniformly on bounded subsets of E_c^*. $\tilde{f} \in \mathcal{E}_\infty$ and the space $\mathcal{E}_\infty$ is exactly the collection of analytic version of members of (E). We reformulate the related results and growth estimates as follows.

Theorem 3.1. [16]

(i) *Let f be in (E). For any $p \geq 1$, let $m_p \in \mathbb{N}$ so that $f \in (E)_{m_p,p}$. Then*

$$\|\tilde{f}\|_{\mathcal{E}_{m_p,p}} \leq c_{m_p} \|f\|_{\mathcal{E}_{m_p,p}}, \tag{3.2}$$

where $c_m = \int_C e^{m|y|_\infty} \mu(dy)$ for any $m \in \mathbb{N}$.

(ii) *Let φ be in $\mathcal{E}_\infty$. For any $p \geq 1$, let $m_p \in \mathbb{N}$ so that $\varphi \in \mathcal{E}_{m_p}(E_{-p})$. Let r be a real number such that $r > \ln 2/(2 \ln \pi - 2 \ln 2)$, and choose $\tilde{m}_p$ be the least integer greater than $\sqrt{m_p\, e}$. Then*

$$\|\varphi\|_{\tilde{m}_p,p-r} \leq \beta_{m_p,p} \|\varphi\|_{\mathcal{E}_{m_p,p}}, \tag{3.3}$$

where $\beta_{m_p,p}$ is a constant depending only on p and m_p.

(iii) *The mapping $\varphi \to \mu * \varphi$ is a homeomorphism from $\mathcal{E}_\infty$ onto $\mathcal{E}_\infty$. In fact, let φ be assumed as above. Then*

$$(1/c_k) \|\varphi\|_{\mathcal{E}_{\tilde{m}_p,p}} \leq \|\mu * \varphi\|_{\mathcal{E}_{m_p,p}} \leq c_{m_p} \|\varphi\|_{\mathcal{E}_{m_p,p}}, \tag{3.4}$$

where k is the least integer greater than e^{m_p}.

(iv) *$\mathcal{E}_\infty \subset (E)$ and $\mathcal{E}_\infty = \widetilde{(E)}$ as vector spaces, where $\widetilde{(E)} = \{\tilde{f} : f \in (E)\}$.*

Proof. The statements (i), (ii), and (iv) follow from [16: Theorem 4.1 and Theorem 4.4]. For the statement (iii), it is obvious that $\|\mu * \varphi\|_{\mathcal{E}_{m_p,p}} \leq c_{m_p} \|\varphi\|_{\mathcal{E}_{m_p,p}}$. On the other hand, take an arbitrary $\psi \in \mathcal{E}_\infty$ and define

$$f(z) = \sum_{n=0}^{\infty} \frac{1}{n!} \int_C D^n \psi(0)(z + i\,y)^n \, \mu(dy), \quad z \in E_c^*.$$

Then $\mu * f = \psi$. For any $p \geq 1$, let $m_p \in \mathbb{N}$ so that $\psi \in \mathcal{E}_{m_p,p}$. Then, by the Cauchy integral formula, $|D^n \psi(0) z^n| \leq e^{m_p n} \|\psi\|_{\mathcal{E}_{m_p,p}} |z|^n_{-p}$ for each $n \in \mathbb{N}$. Thus, for any $z \in E_{-p,c}$,

$$|f(z)| \leq \sum_{n=0}^{\infty} \frac{1}{n!} \int_C |D^n \psi(0)(z + i\,y)^n| \, \mu(dy)$$

$$\leq \|\psi\|_{\mathcal{E}_{m_p,p}} \int_C \exp\{e^{m_p}(|z|_{-p} + |y|_{-p})\} \, \mu(dy).$$

Dividing both sides of the above estimation by $e^{k|z|_{-p}}$, we obtain (3.4). $\qquad \square$

Example 3.2.

(1) For $f_1, f_2, \ldots, f_n \in C'$, define

$$: \widetilde{f}_1 \cdots \widetilde{f}_n : (x) = \int_C \prod_{j=1}^{n} (x + iy, f_j)\, \mu(dy).$$

Then $: \widetilde{f}_1 \cdots \widetilde{f}_n : \ \in L^2(C, \mu)$. It is worth to note that

$$I_n(f_1 \otimes \cdots \otimes f_n) = \ : \widetilde{f}_1 \cdots \widetilde{f}_n : .$$

Moreover, for $\eta_1, \ldots, \eta_n \in E_c$, we have $: \widetilde{\eta}_1 \cdots \widetilde{\eta}_n : \ \in \mathcal{E}_\infty$.

(2) The exponential vector functional $\varepsilon(\eta)$ associated with $\eta \in E_c$ which is given by

$$\varepsilon(\eta) = \exp\left\{ (\cdot, \eta) - \frac{1}{2} \int_0^1 \dot{\eta}(t)^2 \, dt \right\}.$$

Then $\varepsilon(\eta) \in \mathcal{E}_\infty$.

For more examples, we refer the reader to [17].

Remark 3.3.

(a) Identify (E) with $\mathcal{E}_\infty$, then Theorem 3.1 implies that two families of norms $\{\|\cdot\|_{m,p}\}$ and $\{\|\cdot\|_{\mathcal{E}_{m,p}}\}$ are equivalent. In other words, the space $\mathcal{E}_\infty$ is equivalent to the Yan-Meyer space (E).

(b) Let f_{j_i}'s be functions as given in (2.1). Then for any $\varphi \in \mathcal{E}_\infty$, the series

$$\sum_{n=0}^{\infty} \frac{1}{n!} \left\{ \sum_{j_1,\ldots,j_n=1}^{\infty} D^n \mu * \varphi(0)(f_{j_1}, \ldots, f_{j_n}) : \widetilde{f}_{j_1} \cdots \widetilde{f}_{j_n} : (z) \right\},$$

converges to φ in $\mathcal{E}_\infty$ and, for any $p \geq 1$, the series also converges absolutely and uniformly on each bounded set of $E_{-p,c}$. As a consequence, the linear space $\mathcal{P}$ spanned by all cylinder polynomials of the form $\phi((\cdot, \eta_1), \ldots, (\cdot, \eta_n))$ for any $n \in \mathbb{N} \cup \{0\}$ is dense in $\mathcal{E}_\infty$ and in $\mathcal{E}_p$ for $p \geq 1$, where $\phi(x_1, \ldots, x_n)$ is a n-variable polynomial with complex coefficients. However, it is not clear to us if $\mathcal{P}$ is dense in $\mathcal{E}_0$.

Denote the dual spaces of $\mathcal{E}_0$, $\mathcal{E}_p$ $(p \geq 1)$, and $\mathcal{E}_\infty$ respectively by $\mathcal{E}_0^*$, $\mathcal{E}_p^*$, and $\mathcal{E}_\infty^*$ which are topolozied by the weak*-topologies. Then we have the following chain of continuous inclusions:

$$\mathcal{E}_\infty \subset \mathcal{E}_p \subset \mathcal{E}_q \subset \mathcal{E}_0 \subset L^2[C, \mu] \subset \mathcal{E}_0^* \subset \mathcal{E}_q^* \subset \mathcal{E}_p^* \subset \mathcal{E}_\infty^*, \qquad \text{whenever } p \geq q \geq 1.$$

Members of $\mathcal{E}_\infty^*$ will be referred as the generalized Wiener functionals .

Next, let $\langle\!\langle\cdot,\cdot\rangle\!\rangle_0$, $\langle\!\langle\cdot,\cdot\rangle\!\rangle_p$, and $\langle\!\langle\cdot,\cdot\rangle\!\rangle_\infty$ stand for the dual pairing of $\mathcal{E}_0^*$-$\mathcal{E}_0$, $\mathcal{E}_p^*$-$\mathcal{E}_p$, and $\mathcal{E}_\infty^*$-$\mathcal{E}_\infty$, respectively.

Example 3.4. Denote by $L_{\exp}^1[\mathcal{C},\mu]$ the space of all measurable functions f defining on $(\mathcal{C},\mathcal{B}(\mathcal{C}))$ such that

$$\int_{\mathcal{C}} f(x)\, e^{m\,|x|_\infty}\, \mu(dx) \quad \text{for all } m \in \mathbb{N}.$$

By the Fernique theorem, $L_{\exp}^1[\mathcal{C},\mu]$ can be regarded as a subspace of $\mathcal{E}_0^*$ by identifying each $f \in L_{\exp}^1[\mathcal{C},\mu]$ with the functional G_f defined by

$$\langle\!\langle G_f,\varphi\rangle\!\rangle_0 = \int_{\mathcal{C}} f(x)\,\varphi(x)\,\mu(dx),$$

for $\varphi \in \mathcal{E}_0$. Members of $L_{\exp}^1[\mathcal{C},\mu]$ will be called regular generalized Wiener functionals. For more examples we refer the reader to [17].

Denote the dual space of (E) by $(E)^*$ which is endowed with the weak*-topology and let $\langle\!\langle\cdot,\cdot\rangle\!\rangle$ denote the dual pairing of $(E)^*$ and (E).

Definition 3.5. *The S-transform SF of $F \in \mathcal{E}_\infty^*$ is defined as a complex-valued functional on E_c by*

$$SF(\eta) = \langle\!\langle F,\varepsilon(\eta)\rangle\!\rangle_\infty, \qquad \eta \in E_c.$$

We remark that as $F \in L^2(\mathcal{C},\mu)$, $SF = \mu * F$.

Since (E) is dense in $(E)_p$ for any $p \geq 1$, (E) becomes the reduced topological projective limit of $\{(E)_p\}$. Then, by [4: Theorem 6, pp 290], $(E)^* = \cup_{p \geq 1}(E)_p^*$ on which the inductive limit topology is endowed. This implies that if $\widetilde{F} \in (E)^*$, then there exists $p \geq 1$ so that $\widetilde{F} \in (E)_p^*$, the dual of $(E)_p$. Since $(E)_p^* = \cap_{m \geq 1}(E)_{m,p}^*$, we have the following

Proposition 3.6. [16] *Let $\widetilde{F}$ be in $(E)^*$. Then there exists $p \geq 1$ so that for any $m \in \mathbb{N}$,*

$$\sum_{n=0}^\infty \frac{m^{2n}}{(n!)^2}\left\|D^n S\widetilde{F}(0)\right\|_{HS^n(E_p)}^2 < +\infty. \tag{3.5}$$

According to Theorem 3.1, the embedding $j : \mathcal{E}_\infty \to (E)$ is continuous. So that, for any $G \in (E)^*$, $G \circ j \in \mathcal{E}_\infty^*$. Thus $(E)^*$ can be identified as a subspace of $\mathcal{E}_\infty^*$. Conversely, for a fixed $F \in \mathcal{E}_\infty^*$, define a functional $\widetilde{F}$ on (E) by

$$\langle\!\langle \widetilde{F},f\rangle\!\rangle := \langle\!\langle F,\widetilde{f}\rangle\!\rangle_\infty, \quad f \in (E),$$

where $\widetilde{f}$ is the analytic version of f on E_c^*. It follows from Theorem 3.1 that $\widetilde{F} \in (E)^*$.

Recall that a functional G on E_c is analytic if it satisfies the following two conditions:

(A-1) for all $\eta, \phi \in E_c$, the one complex variable mapping $\mathbb{C} \ni \lambda \mapsto G(\eta + \lambda \phi)$ is entire on $\mathbb{C}$; and

(A-2) there exists $p \geq 1$ such that for any $c > 0$, $\sup\{|G(\eta)| : \eta \in E_c \text{ and } |\eta|_p \leq c\} < +\infty$.

The following theorem characterizes the generalized functionals in $\mathcal{E}_\infty^*$ in terms of its S-transforms.

Theorem 3.7. *Let $F \in \mathcal{E}_\infty^*$ be fixed. Then the S-transform SF of F is an analytic function on E_c such that the number*

$$n_{m,-p}(F) := \sum_{n=0}^{\infty} \frac{m^{2n}}{(n!)^2} \|D^n SF(0)\|^2_{HS^n(E_p)}. \tag{3.6}$$

is finite. Conversely, suppose that G is an analytic function defined on E_c and satifies the condition (A-2) *for some $p \geq 1$. Let q be sufficiently large so that $e^2 \cdot \sum_{j=1}^{\infty} \lambda_j^{-2(q-p)} < 1$. Then there exists a unique $F \in \mathcal{E}_\infty^*$ such that $n_{m,-q}(F) < +\infty$ and $SF = G$, where $\lambda_j = (j - (1/2))\pi$ for $j \in \mathbb{N}$.*

Proof. The first assertion follows from Theorem 3.1 (or Remark 3.3), Proposition 3.6, and the following identities:

$$SF(\eta) = \sum_{n=0}^{\infty} \frac{1}{n!} \left\{ \sum_{j_1,\ldots,j_n=1}^{\infty} (\eta, f_{j_1}) \cdots (\eta, f_{j_n}) \langle\!\langle F, : \widetilde{f}_{j_1} \cdots \widetilde{f}_{j_n} : \rangle\!\rangle_\infty \right\}. \tag{3.7}$$

and

$$D^n SF(0)(\eta_1, \ldots, \eta_n) = \sum_{j_1,\ldots,j_n=1}^{\infty} (\eta_1, f_{j_1}) \cdots (\eta_n, f_{j_n}) \langle\!\langle F, : \widetilde{f}_{j_1} \cdots \widetilde{f}_{j_n} : \rangle\!\rangle_\infty. \tag{3.8}$$

To prove the second assertion. Let G is an analytic function on E_c. For any $m \in \mathbb{N}$, since G is locally bounded, there is a positive number M_m so that $|G(\eta)| \leq M_m$ provided that $|\eta|_p \leq m$, $\eta \in E_c$. By the Cauchy integral formula, $\|D^n G(0)\|_{\mathcal{L}^n(E_p)} \leq M_m n^n / m^n$ for each $n \in \mathbb{N}$. Then

$$\sum_{n=0}^{\infty} \frac{m^{2n}}{(n!)^2} \|D^n G(0)\|_{HS^n(E_q)}^2 = \sum_{n=0}^{\infty} \frac{m^{2n}}{(n!)^2} \left\{ \sum_{j_1,\ldots,j_n=1}^{\infty} \lambda_{j_1}^{-2q} \cdots \lambda_{j_n}^{-2q} \right.$$

$$\left. \times \left| D^n G(0)(f_{j_1}, \ldots, f_{j_n}) \right|^2 \right\}$$

$$\leq M_m^2 \sum_{n=0}^{\infty} \frac{n^{2n}}{(n!)^2} \left(\sum_{j=1}^{\infty} \lambda_j^{-2(q-p)} \right)^n$$

$$\leq M_m^2 c^2 \sum_{n=0}^{\infty} \left(e^2 \sum_{j=1}^{\infty} \lambda_j^{-2(q-p)} \right)^n < +\infty,$$

where $c = \sup\{n^{n+(1/2)}/(e^n \, n!) : n \in \mathbb{N}\} < +\infty$ by the Stirling formula (see [8]). Next, for $\varphi \in \mathcal{E}_\infty$, define

$$\langle\!\langle F, \varphi \rangle\!\rangle_\infty = \sum_{n=0}^{\infty} \frac{1}{n!} \left\{ \sum_{j_1,\ldots,j_n=1}^{\infty} D^n \mu * \varphi(0)(f_{j_1}, \ldots, f_{j_n}) \, D^n G(0)(f_{j_1}, \ldots, f_{j_n}) \right\}.$$

By Theorem 3.1, $F \in \mathcal{E}_\infty^*$; and it is clear that $SF = G$. $\qquad\square$

Remark 3.8. Let $\mathcal{F}^\lambda(E_p)$ for $p \in \mathbb{R}$ be the class of analytic functions on $E_{p,c}$ such that

$$\|f\|_{\mathcal{F}^\lambda(E_p)}^2 = \sum_{n=0}^{\infty} \frac{\lambda^n}{n!} \|D^n f(0)\|_{HS^n(E_p)}^2 < +\infty,$$

called the space of Bargmann-Segal analytic functions on $E_{p,c}$ with parameter $\lambda > 0$. Then Lee [13, 15] has shown that

$$\mathcal{A}_{\infty,\lambda} = \cap_{p \geq 1} \mathcal{F}^\lambda(E_{-p}) \subset \mathcal{F}^\lambda(E_{-p}) \subset \mathcal{F}^\lambda(E_{-q}) \subset L^2(\mathcal{C},\mu)$$

$$\subset \mathcal{F}^\lambda(E_q) \subset \mathcal{F}^\lambda(E_p) \subset \cup_{p \geq 1} \mathcal{F}^\lambda(E_p) = S(\mathcal{A}_{\infty,\lambda}^*),$$

$$p \geq q \geq 1,$$

where $\mathcal{A}_{\infty,\lambda}$ is the projective limit of $\{\mathcal{A}_{p,\lambda} : p \geq 1\}$ and $\mathcal{A}_{p,\lambda}$'s are Banach space of analytic functions f on E_c^* with the $\|\cdot\|_{\mathcal{A}_{p,\lambda}}$-norm given by $\|f\|_{\mathcal{A}_{p,\lambda}} = \sup\{|f(z)| e^{-(1/(2\lambda))|z|_{-p}^2} : z \in E_{-p,c}\} < +\infty$; and $\mathcal{A}_{\infty,\lambda}^*$ is the dual of $\mathcal{A}_{\infty,\lambda}$ with the weak*-topology. Comparing with the result in Theorem 3.7, we have the following relations:

1. $\mathcal{E}_\infty \subset \mathcal{A}_{\infty,\lambda} \subset L^2(\mathcal{C},\mu) \subset \mathcal{A}_{\infty,\lambda}^* \subset \mathcal{E}_\infty^*$

2. $S(\mathcal{E}_\infty) = \mathcal{E}_\infty \subset \mathcal{A}_{\infty,\lambda} = S(\mathcal{A}_{\infty,\lambda}) \subset \cup_{p \geq 1} \mathcal{F}^\lambda(E_p) = S(\mathcal{A}_{\infty,\lambda}^*) \subset \mathcal{A}(E_c)$

   $= S(\mathcal{E}_\infty^*),$

where $\mathcal{A}(E_c)$ is the space of all analytic functions on E_c.

Proposition 3.9.

(i) $SF \equiv 0$ iff $F \equiv 0$ for $F \in \mathcal{E}_\infty^*$. Equivalently, the class $\{\varepsilon(\eta) : \eta \in E\}$ is a total subset of $\mathcal{E}_\infty$.

(ii) $\mathcal{E}_\infty$ is a dense subspace of $\mathcal{E}_\infty^*$.

Proof. The first assertion (i) follows from Theorem 3.7. For the second assertion (ii), let $F \in \mathcal{E}_\infty^*$. By Theorem 3.7, there exists $p \geq 1$ so that $n_{m,-p}(F)$ is finite for each $m \in \mathbb{N}$. Let

$$F_k(z) = \sum_{n=0}^{k} \frac{1}{n!} \int_C D^n SF(0)(P_k(z) + i\, P_k(y))^n \, \mu(dy), \quad z \in E_{p,c}, \qquad (3.9)$$

for any $k \in \mathbb{N}$. Then F_k's are all in $\mathcal{E}_\infty$. Take an arbitrary $\varphi \in \mathcal{E}_\infty$. Then $\varphi \in \mathcal{E}_m(E_{-p-r})$ for some $m \in \mathbb{N}$; and by Theorem 3.1, $\|\varphi\|_{\tilde{m},p} < +\infty$, where r and $\tilde{m}$ are defined as in (3.3). Then, by Theorem 3.7 and using the Cauchy-Schwarz inequality,

$$|\langle\!\langle F_k - F, \varphi \rangle\!\rangle_\infty| = |\langle\!\langle F_k - F, \varphi \rangle\!\rangle| \leq \|\varphi\|_{\tilde{m},p} \cdot n_{\tilde{m},-p}(F_k - F) \to 0.$$

Thus, $F_k \to F$ in $\mathcal{E}_\infty^*$ as $k \to \infty$. It implies that $\mathcal{E}_\infty$ is dense in $\mathcal{E}_\infty^*$. $\square$

4 The Clark formula for generalized Wiener functionals

Let $B = \{B(t) : t \in [0,1]\}$ be the standard Brownian motion on $(C, \mathcal{B}(C), \mu)$, where $B(t; x) = x(t)$ for $x \in C$ and $t \in [0,1]$; and let $\mathcal{F}_t$, $t \in [0,1]$, the σ-field generated by $B(u)$, $0 \leq u \leq t$. In [3], Lee and Huang show that the conditional expectation $\mathrm{E}[\varphi | \mathcal{F}_t]$ for $\varphi \in \mathcal{E}_0$ and $t \in [0,1]$ admits an integral representation

$$\mathrm{E}[\varphi | \mathcal{F}_t](z) = \int_C \varphi(P_t(z) + Q_t(y)) \, \mu(dy), \quad z \in C_c, \qquad (4.1)$$

where $P_t(z)(s) = z(t \wedge s)$, $s \in [0,1]$; and $Q_t(z) = z - P_t(z)$. From such a representation, one see that $\mathrm{E}[\varphi | \mathcal{F}_t] \in \mathcal{E}_0$ and that the mapping $T : \varphi \to \mathrm{E}[\varphi | \mathcal{F}_t]$ is continuous on $\mathcal{E}_0$. Since T extends to a self-adjoint operator on $L^2[C, \mu]$. It is natural to define the conditional expectation for a generalized function F in $\mathcal{E}_0^*$ relative to $\mathcal{F}_t$, $t \in [0,1]$, by

$$\langle\!\langle \mathrm{E}[F | \mathcal{F}_t], \varphi \rangle\!\rangle_0 := \langle\!\langle F, \mathrm{E}[\varphi | \mathcal{F}_t] \rangle\!\rangle_0.$$

Consequently, $\mathrm{E}[F | \mathcal{F}_t]$ is in $\mathcal{E}_0^* (\subset \mathcal{E}_\infty^*)$. In this section, we shall prove the Clark formula for generalized functions $F \in \mathcal{E}_0^*$.

140

Note that even for $\eta \in E_c$, $P_t(\eta)$ lies in C_c^* but not necessarily lies in E_c. Therefore, in this section, we restrict our consideration only to the generalized Wiener functionals in $\mathcal{E}_0^*$.

For $F \in \mathcal{E}_0^*$ $(= \cap_{m \in \mathbb{N}} \mathcal{E}_m^*(C))$, let

$$\|F\|_m = \sup\{|\langle\!\langle F, \varphi \rangle\!\rangle_0| : \varphi \in \mathcal{E}_0 \text{ and } \|\varphi\|_{\mathcal{E}_m(C)} = 1\};$$

and, for $\eta \in E_c$, let m_η denote the least integer greater than $\int_0^1 |\ddot{\eta}(s)|\, ds$.

Proposition 4.1. *For $F \in \mathcal{E}_0^*$ and $\eta \in E_c$, the mapping $SF(P_t(\eta))$ is absolutely continuous as a function of one variable with respect to $t \in [0, 1]$.*

Proof. For $0 \le t < u \le 1$ and $z \in C_c$, we observe that

$$|\varepsilon(P_u(h))(z) - \varepsilon(P_t(h))(z)| = \left| \int_C \left(e^{(z+iy,\, P_u(\eta))} - e^{(z+iy,\, P_t(\eta))} \right) \mu(dy) \right|$$

$$\le \left\{ \int_C |z + iy|_\infty \, e^{2\, m_\eta \, |z+iy|_\infty} \, \mu(dy) \right\} \cdot$$

$$\cdot \int_t^u |\ddot{\eta}(s)|\, ds \quad \text{(by Fact 2.1)}$$

$$\le \left\{ \int_C e^{(2\, m_\eta + 1)\, (|z|_\infty + |y|_\infty)} \, \mu(dy) \right\} \int_t^u |\ddot{\eta}(s)|\, ds.$$

It implies that

$$\|\varepsilon(P_u(\eta)) - \varepsilon(P_t(\eta))\|_{\mathcal{E}_{2m_\eta + 1}(C)} \le c_{2m_\eta + 1} \int_t^u |\ddot{\eta}(s)|\, ds, \qquad (4.2)$$

where $c_{2m_\eta + 1}$ is defined as in Theorem 3.1 (i); and thus

$$|SF(P_u(\eta)) - SF(P_t(\eta))| \le c_{2m_\eta + 1} \|F\|_{-2m_\eta - 1} \int_t^u |\ddot{\eta}(s)|\, ds \qquad (4.3)$$

for $0 \le t < u \le 1$. From (4.3) it follows that $t \mapsto SF(P_t(\eta))$ is absolutely continuous. $\qquad \square$

Remark 4.2. Along the line of the proof of Proposition 4.1 with $\int_0^1 |\ddot{\eta}(s)| ds$ being replaced by $\int_0^1 |d\dot{\eta}(s)|$, it is easy to see that Proposition 4.2 remains valid.

For any $\eta \in E_c$, Proposition 4.2 implies that $(d/dt)\, SF(P_t(\eta))$ exists almost everywhere in $(0,1)$ and $(d/dt)\, SF(P_t(\eta))$ is Lebesgue integrable so that

$$SF(P_b(\eta)) - SF(P_a(\eta)) = \int_a^b \frac{d}{dt}\, SF(P_t(\eta))\, dt, \quad 0 \le a < b \le 1.$$

In particular,

$$SF(\eta) = \langle\!\langle F, 1 \rangle\!\rangle_\infty + \int_0^1 \frac{d}{dt}\, SF(P_t(\eta))\, dt. \tag{4.4}$$

For any $t \in [0,1]$, let $T_F(t,\cdot)$ be a complex-valued mapping on E_c given by

$$T_F(t,\eta) = \frac{d}{dt}\, SF(P_t(\eta)), \quad \eta \in E_c.$$

Proposition 4.3. *For $F \in \mathcal{E}_0^*$ and $t \in [0,1]$, $T_F(t,\cdot)$ is analytic on E_c.*

Proof. First of all, we observe that for a sufficiently small $\epsilon > 0$, it follows from (4.3) that

$$\left| \frac{SF(P_{t+\epsilon}(\eta)) - SF(P_t(\eta))}{\epsilon} \right| = |\epsilon|^{-1} \cdot c_{2m_\eta + 1} \cdot \|F\|_{-2m_\eta - 1} \cdot \int_t^{t+\epsilon} |\ddot{\eta}(s)|\, ds$$

$$\le c_{2m_\eta + 1} \cdot \|F\|_{-2m_\eta - 1} \cdot |\ddot{\eta}|_\infty. \tag{4.5}$$

Let η, ϕ be arbitrarily given in E_c. By (4.5) and the dominated convergence theorem, we obtain that

$$\int_\Gamma T_F(t, \eta + \lambda \phi)\, d\lambda$$

$$= \lim_{\epsilon \to 0} \epsilon^{-1} \int_\Gamma \left(SF(P_{t+\epsilon}(\eta + \lambda \phi)) - SF(P_t(\eta + \lambda \phi)) \right) d\lambda$$

$$= \lim_{\epsilon \to 0} \epsilon^{-1} \int_\Gamma \left(S(\mathrm{E}[\,F\,|\,\mathcal{F}_{t+\epsilon}])(\eta + \lambda \phi) - S(\mathrm{E}[\,F\,|\,\mathcal{F}_t])(\eta + \lambda \phi) \right) d\lambda = 0$$

because of the analyticity of SG for $G \in \mathcal{E}_\infty^*$ on E_c, where Γ is a simple closed curve on the complex number plane. By the Morera theorem (see [21]), $\mathbb{C} \ni \lambda \mapsto T_F(t, \eta + \lambda \phi)$ is entire on $\mathbb{C}$. Letting $\epsilon \to 0$ on each side of (4.5), we see that

$$|T_F(t, \eta)| \le c_{2m_\eta + 1} \cdot \|F\|_{-2m_\eta - 1} \cdot |\ddot{\eta}(t)|$$

$$= c_{2m_\eta + 1} \cdot \|F\|_{-2m_\eta - 1} \cdot \left| \int_0^t \eta^{(3)}(s)\, ds \right|$$

$$\le c_{2m_\eta + 1} \cdot \|F\|_{-2m_\eta - 1} \cdot |\eta|_2 \tag{4.6}$$

by Proposition 2.3. Then $T_F(t, \cdot)$ is locally bounded on E_c. Therefore, $T_F(t, \cdot)$ is analytic on E_c. $\qquad\square$

Applying Theorem 3.7 to Proposition 4.3, there exists a unique generalized function $K_F(t)$ in $\mathcal{E}_\infty^*$ such that $SK_F(t)(\eta) = T_F(t, \eta)$ for each $\eta \in E_c$. By using (4.6), we follow the argument in the proof of Theorem 3.7 to see that for any $m > 0$,

$$n_{m,-q}(K_F(t)) \leq C(m, q, F) < +\infty, \tag{4.7}$$

provided that q is large enough so that $e^2 \sum_{j=1}^\infty \lambda_j^{-2(q-2)} < +\infty$, where $C(m, q, F)$ is a constant, independent of the choice of t, given by

$$C(m, q, F) = M_m^2 \, c^2 \sum_{n=0}^\infty \left(e^2 \sum_{j=1}^\infty \lambda_j^{-2(q-2)} \right)^n,$$

where $M_m = m \, c_{2m+1} \, \|F\|_{-2m-1}$ and $c = \sup\{n^{n+(1/2)}/(e^n \, n!) : n \in \mathbb{N}\}$.

Definition 4.4. *Let G be a $\mathcal{E}_\infty^*$-valued function on $[0, 1]$. We define the integral $\int_0^1 G(t)\, dt$ as*

$$\lim_{\|\Delta\| \to 0} \sum_{i=1}^n (t_i - t_{i-1})\, G(t_i') \quad \text{in } \mathcal{E}_\infty^*,$$

provided that the limit exists, where $\Delta = \{0 = t_0 < t_1 < \cdots < t_n = 1\}$ is any partition of $[0, 1]$, $\|\Delta\|$ is the mesh of Δ, and t_i''s are arbitrarily taken over $[t_i, t_{i-1}]$.

By Theorem 3.1 and the inequality (4.7), the integral $\int_0^1 K_F(t)\, dt$ exists in $\mathcal{E}_\infty^*$. Moreover, for any $\varphi \in \mathcal{E}_\infty$,

$$\left\langle\!\!\left\langle \int_0^1 K_F(t)\, dt, \; \varphi \right\rangle\!\!\right\rangle_\infty = \int_0^1 \langle\!\langle K_F(t), \varphi \rangle\!\rangle_\infty \, dt.$$

It now follows from (4.4) that we obtain

$$SF(\eta) = \langle\!\langle F, 1 \rangle\!\rangle_\infty + S\left(\int_0^1 K_F(t)\, dt \right)(\eta) \quad \text{for } F \in \mathcal{E}_0^* \text{ and } \eta \in E_c.$$

By Proposition 3.6, we obtain the Clark formula for a generalized function $F \in \mathcal{E}_0^*$ below.

Theorem 4.5. *Let F be in $\mathcal{E}_0^*$. Then there exists a unique $K_F(t) \in \mathcal{E}_\infty^*$ so that*

$$S(K_F(t))(\eta) = \frac{d}{dt} S(\mathrm{E}[\,F\,|\,\mathcal{F}_t\,])(\eta)$$

for any $t \in [0,1]$ and $\eta \in E_c$. Moreover, the integral $\int_0^1 K_F(t)\, dt$ exists in $\mathcal{E}_\infty^*$ and

$$F = \langle\!\langle F, 1 \rangle\!\rangle_\infty + \int_0^1 K_F(t)\, dt.$$

Corollary 4.6. *For $\varphi \in \mathcal{E}_1$,*

$$\varphi(x) = \mathrm{E}[\,\varphi\,] + \int_0^1 \mathrm{E}[\,D\varphi(x)\,1_{[t,1]}\,|\,\mathcal{F}_t\,]\,dB(t;\,x) \qquad \mu\text{-almost all } x \text{ on } C,$$

where the integral is the Wiener-Itô stochastic integral.

Proof. First, we note that $D\varphi(\cdot)\,1_{[t,1]} \in \mathcal{E}_1$ for $t \in [0,1]$. Then the conditional expectation $\mathrm{E}[\,D\varphi(\cdot)\,1_{[t,1]}\,|\,\mathcal{F}_t\,]$ lies in $L^2(C,\mu)$. By the Kubo-Takenaka formula (see [17]),

$$S\left(\int_0^1 \mathrm{E}[\,D\varphi(\cdot)\,1_{[t,1]}\,|\,\mathcal{F}_t\,]\,dB(t) \right)(\eta) \quad (\eta \in E_c)$$

$$= \int_0^1 \langle\!\langle \mathrm{E}[\,D\varphi(\cdot)\,1_{[t,1]}\,|\,\mathcal{F}_t\,],\, D\varepsilon(\eta)(\cdot)\,1_{[t,1]} \rangle\!\rangle_\infty\, dt. \tag{4.8}$$

By a direct computation, the equality (4.8) becomes

$$\int_0^1 \dot\eta(t)\, S\big(\mathrm{E}[\,D\varphi(\cdot)\,1_{[t,1]}\,|\,\mathcal{F}_t\,]\big)(\eta)\, dt = \int_0^1 \dot\eta(t)\, S\big(D\varphi(\cdot)\,1_{[t,1]}\big)(P_t(\eta))\, dt$$

$$= \int_0^1 \frac{d}{dt} S\varphi(P_t(\eta))\, dt.$$

Applying Theorem 4.5 we see that

$$\int_0^1 \mathrm{E}[\,D\varphi(\cdot)\,1_{[t,1]}\,|\,\mathcal{F}_t\,]\,dB(t) = \int_0^1 K_F(t)\, dt = \varphi - \mathrm{E}[\,\varphi\,].$$

$$\square$$

Remark 4.7.

(a) For any $F \in \mathcal{E}_0^*$, the $\mathcal{E}_\infty^*$-valued function $[0,1] \ni t \to K_F(t).$ is called the Clark kernel function on $[0,1]$ associated with F.

(b) The formula in Corollary 4.6 can be extended to all the generalized functions in $\mathcal{E}_0^*$. Since the arguments are more involved, we shall prove this result in our forthcoming paper [18].

References

1. J. M. C. Clark: The representation of functionals of Brownian motion by stochastic integrals, *Ann. Math. Stat.* 41 (1970), 1281-1295; 42 (1971), 1778
2. M. de Faria, M. J. Oliveira, and L. Streit: A generalized Clark-Ocone formula, Preprint, 1999.
3. H.-C. Huang and Y.-J. Lee: Conditional expectation of generalized Wiener functionals, 2001, Preprint
4. G. Köthe: *Topological Vector Space I*, Springer-Verlag, New York/ Heidelberg/ Berlin, 1966
5. I. Karatzas, D. Ocone, and J. Li: An extension of Clark's formula, *Stochastics and Stochastics Reports* 37 (1991), 127-131
6. H. Korezlioglu and A. S. Üstünel: New class of distributions on Wiener spaces, *Lecture Notes in Math.* Vol. 1444, Springer-Verlag, 1990
7. H.-H. Kuo: *Gaussian Measures in Banach Spaces*, Lectures Notes in Math. Vol. 463, 1975
8. H.-H. Kuo: *White Noise Distribution Theory*, CRC Press, 1996
9. Y.-J. Lee: Application of Fourier-Wiener transform to differential equations on infinite dimensional spaces I, *Trans. Amer. Math. Soc.* 262 (1980), 218-236
10. Y.-J. Lee: Sharp inequalities and regularity of heat semigroup on infinite dimensional space, *J. Funct. Anal.* 71 (1987), 69-87
11. Y.-J. Lee: Generalized functions on infinite dimensional spaces and its application to white noise calculus, *J. Funct. Anal.* 82 (1989), 429-464
12. Y.-J. Lee: On the convergence of Wiener-Itô decomposition, *Bull. Inst. Math. Acad. Sinica* 17 (1989), 305-312
13. Y.-J. Lee: Analytic version of test functionals, Fourier transform and a characterization of measures in white noise calculus, *J. Funct. Anal.* 100 (1991), 359-380
14. Y.-J. Lee: Positive generalized functions on infinite dimensional spaces, In *"Stochastic Process, a Festschrift in Honour of Gopinath Kallianpur"*, Springer-Verlag, 1993, 225-234
15. Y.-J. Lee: Convergence of Fock expansion and transformations of Brownian functionals, in *"Functional Analysis and Global Analysis"*, Edited by T. Sunada and P.-W. Sy, Springer-Verlag, 1997, 142-156

16. Y.-J. Lee: Integral representation of second quantization and its application to white noise analysis, *J. Funct. Anal.* 133 (1995), 253-276
17. Y.-J. Lee: Generalized white noise functionals on classical Wiener space, *J. Korean Math. Soc.* 35 No.3 (1998), 613-635
18. Y.-J. Lee and H.-H. Shih: Generalized Clark formula on the classical Wiener space, 2001, in preparation
19. P. A. Meyer and J.-A. Yan: Les "fonctions caracteéristiques" des distributionc sur l'espace de Wiener, Sém. Probab. XXV, *Lecture Notes in Math.* 1485 (1991), 61-78
20. D. Ocone: Malliavin's calculus and stochastic integral representations of functionals of diffusion process, *Stochastics.* 12 (1984), 161-185
21. W. Rudin: *"Real and Complex Analysis"*, 3d ed., McGraw-Hill Book Company, New York, 1987
22. A. S. Üstünel: Representations of distributions on Wiener space and stochastic calculus of variations, *J. Funct. Anal.* 70 (1987), 126-139
23. S. Watanabe: *Lectures on Stochastic Differential Equations and Malliavin Calculus*, Tata Inst. of Fundamental Research, Bombay, 1984

Quantum Information IV (pp. 147–176)
Eds. T. Hida and K. Saitô
© 2002 World Scientific Publishing Co.

INVERSE S-TRANSFORM, WICK PRODUCT AND OVERCOMPLETENESS OF EXPONENTIAL VECTORS

NOBUAKI OBATA

Graduate School of Information Sciences
Tohoku University
Sendai 980-8579 Japan
E-mail: obata@math.is.tohoku.ac.jp

By means of distribution theory over a complex Gaussian space we formulate a coherent state representation of a white noise function and a diagonal coherent state representation of a white noise operator. An inversion formula for the S-transform (the Segal–Bargmann transform) is written down in terms of coherent state representation. As a kernel distribution the S-transform is characterized by a diagonal coherent state representation of a Wick multiplication operator. The null kernel of the coherent state representation is described in terms of holomorphic distributions and the overcompleteness of the exponential vectors is explained clearly.

Introduction

We start with the real Gelfand triple:

$$E \equiv \mathcal{S}(\mathbf{R}) \subset H \equiv L^2(\mathbf{R}) \subset E^* \equiv \mathcal{S}'(\mathbf{R}).$$

The canonical bilinear form on $E^* \times E$, which is compatible to the inner product of H, is denoted by $\langle \cdot, \cdot \rangle$ and the norm of H by $|\cdot|_0$. The Gaussian measure on E^* with variance σ^2, denoted by μ_{σ^2}, is uniquely determined by the characteristic function:

$$\exp\left\{-\frac{\sigma^2}{2}\,|\xi|_0^2\right\} = \int_{E^*} e^{i\langle x, \xi \rangle}\, \mu_{\sigma^2}(dx), \quad \xi \in E.$$

The *real Gaussian space* is by definition the space E^* equipped with $\mu = \mu_1$ and the *complex Gaussian space* is $E_{\mathbf{C}}^* = E^* + iE^*$ equipped with

$$\nu(dz) = \mu_{1/2}(dx)\mu_{1/2}(dy), \quad z = x + iy. \tag{1}$$

Both are probability spaces.

In Gaussian functional analysis an important role is played by an *exponential vector* (or also called a *coherent vector*) defined by

$$\phi_\xi(x) = e^{\langle x, \xi \rangle - \langle \xi, \xi \rangle / 2}, \qquad \xi \in E_{\mathbf{C}}, \quad x \in E^*, \tag{2}$$

where $\langle \cdot, \cdot \rangle$ is the $\mathbf{C}$-bilinear form on $E_{\mathbf{C}}^* \times E_{\mathbf{C}}$. It is known that $\{\phi_\xi\,;\,\xi \in E_{\mathbf{C}}\}$ is linearly independent and spans a dense subspace of $L^2(E^*, \mu)$. If $\xi \in E_{\mathbf{C}}$ in

(2) is replaced with $z \in E_{\mathbf{C}}^*$, the function ϕ_z is no longer an ordinary function but still can be formulated as a distribution on E^*. Then, one may be led naturally to a linear combination of such exponential vectors in an integral form:

$$\Phi = \int_{E_{\mathbf{C}}^*} \rho(z)\phi_z\, \nu(dz) \tag{3}$$

called a *coherent state representation*. In fact, it is more natural and important to consider operators rather than functions for functions constitute a particular class of operators. With each $z \in E_{\mathbf{C}}^*$ we associate an operator Q_z defined by the formula:

$$Q_z\phi = \langle\!\langle \phi_{\bar{z}},\, \phi \rangle\!\rangle\, \phi_z.$$

Apart from normalization Q_z is an analogue of one-dimensional projection on $L^2(E^*, \mu)$, but a precise meaning as a white noise operator is given only through distribution theory. It is then natural to consider an operator version of (3), that is,

$$\Xi = \int_{E_{\mathbf{C}}^*} \rho(z)Q_z\, \nu(dz) \tag{4}$$

called a *diagonal coherent state representation*. This idea traces back to Klauder[12] in the case of finite dimension and has been extensively studied[1,13]. Rigorous definitions of (3) and (4) based on the so-called white noise distribution theory[16,18] was first given by Obata[20,21].

To be more precise, we need a white noise triple:

$$\mathcal{W} \subset \Gamma(L^2(\mathbf{R})) \cong L^2(E^*, \mu) \subset \mathcal{W}^*, \tag{5}$$

where $\mathcal{W}$ and $\mathcal{W}^*$ are, respectively, the spaces of test white noise functions and of white noise distributions on the real Gaussian space (E^*, μ). In this paper we adopt the CKS-space[6]. Then, in view of (5) and $L^2(E_{\mathbf{C}}^*, \nu) \cong L^2(E^*, \mu_{1/2}) \otimes L^2(E^*, \mu_{1/2})$, which follows from (1), a *CKS-space over the complex Gaussian space* is constructed:

$$\mathcal{D} \subset L^2(E_{\mathbf{C}}^*, \nu) \subset \mathcal{D}^*.$$

It will be shown that exressions (3) and (4) are justified for any $\rho \in \mathcal{D}^*$ and represent $\Phi \in \mathcal{W}^*$ and $\Xi \in \mathcal{L}(\mathcal{W}, \mathcal{W}^*)$, respectively (Lemmas 3.4 and 4.2).

Notice that representation (3) is not unique due to the famous overcompleteness of exponential vectors; for example, for $\xi \in E_{\mathbf{C}}$ we have

$$\phi_\xi = \int_{E_{\mathbf{C}}^*} e^{\langle \bar{z}, \xi \rangle} \phi_z \nu(dz) = \int_{E_{\mathbf{C}}^*} e^{\langle \bar{z}, \xi \rangle + \langle z, \xi \rangle - \langle \xi, \xi \rangle} \phi_z \nu(dz). \tag{6}$$

One of the main purposes of this paper is to clarify this phenomenon. A key role is played by the S-transform[14] which is a white noise extension of the famous Segal–Bargmann transform[8].

Let $L^2(E_{\mathbf{C}}^*, \nu)_{\mathrm{HOL}}$ denote the space of holomorphic L^2-functions on the complex Gaussian space $(E_{\mathbf{C}}^*, \nu)$. The Segal–Bargmann transform is a unitary isomorphism from $L^2(E^*, \mu)$ onto $L^2(E_{\mathbf{C}}^*, \nu)_{\mathrm{HOL}}$ uniquely defined by the correspondence:

$$\phi_\xi(x) = e^{\langle x, \xi\rangle - \langle \xi, \xi\rangle/2} \quad \leftrightarrow \quad \epsilon_\xi(z) = e^{\langle z, \xi\rangle}, \qquad \xi \in E_{\mathbf{C}}.$$

On the other hand, the S-transform is defined for $\Phi \in \mathcal{W}^*$ by

$$S\Phi(\xi) = \langle\!\langle \Phi, \phi_\xi \rangle\!\rangle, \qquad \xi \in E_{\mathbf{C}}.$$

If $\Phi \in L^2(E^*, \mu)$, the S-transform $S\Phi$ is extended to a holomorphic L^2-function on $(E_{\mathbf{C}}^*, \nu)$ and coincides with the Segal–Bargmann transform. Moreover, for $\Phi \in \mathcal{W}^*$ the S-transform $S\Phi$ is viewed as a holomorphic distribution on $E_{\mathbf{C}}^*$ and one comes to the following

Theorem 3.3 *The S-transform extends the Segal–Bargmann transform and is a topological isomorphism from $\mathcal{W}^*$ onto $\mathcal{D}_{\mathrm{HOL}}^*$.*

The inverse S-transform is given by the coherent state representation.

Theorem 3.5 *For any $\Phi \in \mathcal{W}^*$, the S-transform $S\Phi$ being regarded as a holomorphic distribution on the complex Gaussian space, it holds that*

$$\Phi = \int_{E_{\mathbf{C}}^*} S\Phi(\bar{z})\phi_z \nu(dz). \tag{7}$$

In particular, every white noise distribution Φ admits a coherent state representation.

However, since a coherent state representation is not unique due to the overcompleteness of exponential vectors, there arises a question of characterization of the S-transform as a kernel in (7). To answer this question, we focus on uniqueness of the diagonal coherent state representation for white noise operators, that is,

Theorem 4.3 *Every white noise operator $\Xi \in \mathcal{L}(\mathcal{W}, \mathcal{W}^*)$ admits a unique diagonal coherent state representation as in (4) with $\rho \in \mathcal{D}^*$.*

Applying this to a Wick multiplication operator, we come to the following

Theorem 5.2 *Let W_Φ be the Wick multiplication operator by $\Phi \in \mathcal{W}^*$. Then*

$$W_\Phi = \int_{E_{\mathbf{C}}^*} S\Phi(\bar{z})Q_z \nu(dz). \tag{8}$$

Thus the S-transform $S\Phi$ is characterized as a unique kernel distribution of W_Φ. The inverse S-transform is again obtained from (8) by using simple relations $W_\Phi\phi_0 = \Phi$ and $Q_z\phi_0 = \phi_z$. Now, the first integral in (6) corresponds to the Wick multiplication:

$$W_{\phi_\xi} = \int_{E_{\mathbf{C}}^*} e^{\langle \bar{z}, \xi \rangle} Q_z \nu(dz),$$

while the second identity:

$$\phi_\xi = \int_{E_{\mathbf{C}}^*} e^{\langle \bar{z}, \xi \rangle + \langle z, \xi \rangle - \langle \xi, \xi \rangle} Q_z \nu(dz)$$

gives the diagonal coherent state representation of the pointwise multiplication operator by ϕ_ξ.

Finally, we investigate the null kernel of coherent state representations. Let $\mathcal{D}_{\mathrm{AH}}$ be the space of all anti-holomorphic test functions and $\mathcal{D}_{\mathrm{AH}}^\perp$ denote the space of all $\rho \in \mathcal{D}^*$ which annihilate $\mathcal{D}_{\mathrm{AH}}$, that is, all $\rho \in \mathcal{D}^*$ such that $\langle\!\langle \rho, \phi \rangle\!\rangle_\nu = 0$ for all $\phi \in \mathcal{D}_{\mathrm{HOL}}$. (By our convention $\langle\!\langle \cdot, \cdot \rangle\!\rangle_\nu$ stands for the canonical $\mathbf{C}$-bilinear form on $\mathcal{D}^* \times \mathcal{D}$.)

Theorem 5.4 *Let $\Phi \in \mathcal{W}^*$. Then $\rho \in \mathcal{D}^*$ is a kernel of its coherent state representation:*

$$\Phi = \int_{E_{\mathbf{C}}^*} \rho(z)\phi_z \, \nu(dz)$$

if and only if

$$\rho = \rho_1 + \rho_2, \qquad \rho_1 \in \mathcal{D}_{\mathrm{AH}}^*, \quad \rho_2 \in \mathcal{D}_{\mathrm{AH}}^\perp,$$

with $\rho_1(z) = S\Phi(\bar{z})$.

We have thus seen that complex white noise theory established in this paper sheds some light on questions concerning the overcompleteness of exponential vectors. In particular, it seems worthwhile to note that the S-transform is characterized in terms of coherent state representation.

Complex white noise dates back to the early stages of white noise theory and has been studied from different aspects. Among others, a white noise approach to the Segal–Bargmann transform, discussed by Kubo–Yokoi[15] and Yokoi[26], has a close relation to the present work. While, a new direction of characterization theorems is proposed by Ji–Obata[10] where we enjoy an interesting encounter with theory of infinite holomorphy[11,24].

1 White Noise Functions

1.1 Weighted Fock space

Let $\mathcal{H}$ be a (real or complex) Hilbert space. For $n \geq 1$ the n-fold symmetric tensor power of $\mathcal{H}$ is denoted by $\mathcal{H}^{\hat{\otimes}n}$ and by definition $\mathcal{H}^{\hat{\otimes}0}$ is one dimensional. Given a sequence $\alpha = \{\alpha(n)\}_{n=0}^{\infty}$ of positive numbers we put

$$\Gamma_{\alpha}(\mathcal{H}) = \left\{ \phi = (f_n)_{n=0}^{\infty} \, ; \, f_n \in \mathcal{H}^{\hat{\otimes}n}, \ \|\phi\|^2 \equiv \sum_{n=0}^{\infty} n! \, \alpha(n)|f_n|^2 < \infty \right\}.$$

In an obvious manner $\Gamma_{\alpha}(\mathcal{H})$ becomes a Hilbert space and is called the *weighted Fock space* over $\mathcal{H}$. The *Boson Fock space* is by definition the special case of $\alpha(n) \equiv 1$ and is denoted by $\Gamma(\mathcal{H})$.

We need some general notion for positive sequences. A positive sequence $\alpha = \{\alpha(n)\}_{n=0}^{\infty}$ is called *log-concave* if

$$\alpha(n)\alpha(n+2) \leq \alpha(n+1)^2, \qquad n = 0, 1, 2, \dots.$$

Two positive sequences $\alpha = \{\alpha(n)\}_{n=0}^{\infty}$ and $\beta = \{\beta(n)\}_{n=0}^{\infty}$ are called *equivalent* if there exist positive constants $K_1, K_2, M_1, M_2 > 0$ such that

$$K_1 M_1^n \alpha(n) \leq \beta(n) \leq K_2 M_2^n \alpha(n), \qquad n = 0, 1, 2, \dots.$$

From now on we fix a weight sequence $\alpha = \{\alpha(n)\}_{n=0}^{\infty}$ satisfying the following conditions:

(A1) $\alpha(0) = 1$ and $\displaystyle\inf_{n \geq 0} \sigma^n \alpha(n) > 0$ for some $\sigma \geq 1$;

(A2) $\displaystyle\lim_{n \to \infty} \left\{ \frac{\alpha(n)}{n!} \right\}^{1/n} = 0$;

(A3) α is equivalent to a positive sequence $\gamma = \{\gamma(n)\}$ such that $\{\gamma(n)/n!\}$ is log-concave;

(A4) α is equivalent to another positive sequence $\gamma = \{\gamma(n)\}$ such that $\{(n!\gamma(n))^{-1}\}$ is log-concave.

For example, $(n!)^{\beta}$ with $0 \leq \beta < 1$ and the Bell numbers of order k satisfy the above conditions.

The exponential generating functions of α and $1/\alpha$ are defined by

$$G_{\alpha}(t) = \sum_{n=0}^{\infty} \frac{\alpha(n)}{n!} t^n, \qquad G_{1/\alpha}(t) = \sum_{n=0}^{\infty} \frac{1}{n!\alpha(n)} t^n,$$

respectively. Both are entire holomorphic functions by (A1) and (A2). We next define

$$\widetilde{G}_\alpha(t) \equiv \sum_{n=0}^{\infty} t^n \frac{n^{2n}}{n!\alpha(n)} \left\{ \inf_{s>0} \frac{G_\alpha(s)}{s^n} \right\},$$

$$\widetilde{G}_{1/\alpha}(t) \equiv \sum_{n=0}^{\infty} t^n \frac{n^{2n}\alpha(n)}{n!} \left\{ \inf_{s>0} \frac{G_{1/\alpha}(s)}{s^n} \right\}.$$

It is known[2] that (A3) and (A4) are necessary and sufficient conditions respectively for $\widetilde{G}_\alpha$ and for $\widetilde{G}_{1/\alpha}$ to have positive radiai of convergence. These functions will play a crucial role in norm estimates. Moreover, the next fact is known[2].

Lemma 1.1 *Let $\alpha = \{\alpha(n)\}$ be as above.*

(1) *There exists a constant $C_1 > 0$ such that*

$$\alpha(n)\alpha(m) \leq C_1^{n+m}\alpha(n+m), \qquad n, m = 0, 1, 2, \ldots.$$

(2) *There exists a constant $C_2 > 0$ such that*

$$\alpha(n+m) \leq C_2^{n+m}\alpha(n)\alpha(m), \qquad n, m = 0, 1, 2, \ldots.$$

(3) *There exists a constant $C_3 > 0$ such that*

$$\alpha(n) \leq C_3^m\alpha(m), \qquad n \leq m.$$

Then, by a simple calculation we have

Proposition 1.2 *Let $\alpha = \{\alpha(n)\}$ be as above and $G_\alpha(t)$ the exponential generating function. Then, for $s, t \geq 0$ we have:*

(1) $G_\alpha(0) = 1$ *and* $G_\alpha(s) \leq G_\alpha(t)$ *for* $s \leq t$.

(2) $G_\alpha(s)G_\alpha(t) \leq G_\alpha(C_1(s+t))$.

(3) $G_\alpha(s+t) \leq G_\alpha(C_2 s)G_\alpha(C_2 t)$.

(4) $e^s G_\alpha(t) \leq G_\alpha(C_3(s+t))$.

(5) $e^t \leq M G_\alpha(\sigma t)$, *where σ is introduced in (A1) and $M^{-1} = \inf_{n \geq 0} \sigma^n \alpha(n)$.*

1.2 CKS-Space

As usual, we start with the real Gelfand triple:

$$E \equiv \mathcal{S}(\mathbf{R}) \subset H \equiv L^2(\mathbf{R}) \subset E^* \equiv \mathcal{S}'(\mathbf{R}), \tag{9}$$

where the canonical bilinear form on $E^* \times E$ is denoted by $\langle \cdot, \cdot \rangle$ and the norm of $L^2(\mathbf{R})$ by $|\cdot|_0$. For $p \geq 0$ let $E_{\pm p}$ be the Hilbert space obtained by completing $E = \mathcal{S}(\mathbf{R})$ with respect to the norm $|\xi|_{\pm p} = |A^{\pm p}\xi|_0$, where $A = 1 + t^2 - d^2/dt^2$. We then have

$$E \cong \operatorname*{proj\,lim}_{p \to \infty} E_p, \qquad E^* \cong \operatorname*{ind\,lim}_{p \to \infty} E_{-p}.$$

In general, for a real vector space $\mathfrak{X}$ the complexification is denoted by $\mathfrak{X}_{\mathbf{C}}$. For notational convenience, the $\mathbf{C}$-bilinear extension on $E_{\mathbf{C}}^* \times E_{\mathbf{C}}$ is denoted by the same symbol. It is then noted that $|\xi|_0^2 = \langle \bar{\xi}, \xi \rangle$ for $\xi \in H_{\mathbf{C}}$.

Let $\Gamma_\alpha(E_p)$ be the weighted Fock space over E_p. Then

$$\mathcal{W} = \Gamma_\alpha(E) = \operatorname*{proj\,lim}_{p \to \infty} \Gamma_\alpha(E_p)$$

becomes a nuclear Fréchet space and we obtain a Gelfand triple:

$$\mathcal{W} = \Gamma_\alpha(E) \subset \Gamma(H_{\mathbf{C}}) \subset \Gamma_\alpha(E)^* = \mathcal{W}^*, \tag{10}$$

which is called the *Cochran–Kuo–Sengupta space*[6] or the *CKS-space* for short. The topology of $\mathcal{W}$ is defined by the family of norms:

$$\|\phi\|_{p,+}^2 = \sum_{n=0}^{\infty} n!\,\alpha(n)|f_n|_p^2, \qquad \phi = (f_n), \quad p \geq 0.$$

As a consequence of standard argument,

$$\Gamma_\alpha(E)^* \cong \operatorname*{ind\,lim}_{p \to \infty} \Gamma_{\alpha^{-1}}(E_{-p}),$$

where $\Gamma_\alpha(E)^*$ carries the strong dual topology and $\cong$ stands for a topological linear isomorphism. The canonical $\mathbf{C}$-bilinear form on $\mathcal{W}^* \times \mathcal{W}$ is denoted by $\langle\!\langle \cdot, \cdot \rangle\!\rangle$. Then

$$\langle\!\langle \Phi, \phi \rangle\!\rangle = \sum_{n=0}^{\infty} n!\,\langle F_n, f_n \rangle, \qquad \Phi = (F_n) \in \mathcal{W}^*, \quad \phi = (f_n) \in \mathcal{W},$$

and it holds that

$$|\langle\!\langle \Phi, \phi \rangle\!\rangle| \leq \|\Phi\|_{-p,-}\|\phi\|_{p,+},$$

where

$$\|\Phi\|^2_{-p,-} = \sum_{n=0}^{\infty} \frac{n!}{\alpha(n)} \, |F_n|^2_{-p}, \qquad \Phi = (F_n) \in \mathcal{W}^*.$$

Conditions (A1)–(A4) are sorted out by Asai–Kubo–Kuo[2] from many similar ones that have been introduced to keep "nice" properties of a CKS-space. On the other hand, it is also possible to start with a generating function G_α or another function controlling growth rate. This reversed approach is concise and useful for some questions[2,7,11], however, we prefer to the explicit description for our later calculation.

1.3 *Wiener–Itô–Segal Isomorphism and Exponential Vectors*

Let μ_{σ^2} be the Gaussian measure with variance σ^2 defined by the characteristic function:

$$e^{-\sigma^2 |\xi|_0^2/2} = \int_{E^*} e^{i\langle x, \xi\rangle} \mu_{\sigma^2}(dx), \qquad \xi \in E. \tag{11}$$

We put $\mu = \mu_1$ for simplicity and let $L^2(E^*, \mu)$ denote the Hilbert space of $\mathbf{C}$-valued L^2-functions on E^*. The celebrated Wiener–Itô–Segal isomorphism is a unitary isomorphism between $L^2(E^*, \mu)$ and $\Gamma(H_{\mathbf{C}})$ uniquely determined by the correspondence

$$\phi_\xi(x) = e^{\langle x, \xi\rangle - \langle \xi, \xi\rangle/2} \quad \leftrightarrow \quad \left(1, \xi, \frac{\xi^{\otimes 2}}{2!}, \ldots, \frac{\xi^{\otimes n}}{n!}, \ldots\right), \quad \xi \in E_{\mathbf{C}}. \tag{12}$$

The same symbol ϕ_ξ is used for the right hand side. We call ϕ_ξ an *exponential vector* or a *coherent vector*. Note that

$$\langle\!\langle \phi_\xi, \phi_\eta \rangle\!\rangle = e^{\langle \xi, \eta\rangle}, \qquad \xi, \eta \in E_{\mathbf{C}}.$$

Proposition 1.3 $\{\phi_\xi \, ; \, \xi \in E_{\mathbf{C}}\}$ *spans a dense subspace of* $\mathcal{W} = \Gamma_\alpha(E_{\mathbf{C}})$.

The proof is well known.

2 Complex White Noise

2.1 *Complex Gaussian Space*

We define a probability measure ν on $E_{\mathbf{C}}^* = E^* + iE^*$ by

$$\nu(dz) = \mu_{1/2}(dx)\mu_{1/2}(dy), \qquad z = x + iy \in E_{\mathbf{C}}^*,$$

where $\mu_{1/2}$ is the Gaussian measure on E^* with variance $\sigma^2 = 1/2$, see (11). After Hida[9] the probability space $(E^*_{\mathbf{C}}, \nu)$ is called the *complex Gaussian space*. We write $\bar{z} = x - iy$ for $z = x + iy \in E^* + iE^*$. Here are basic formulae:

$$\int_{E^*_{\mathbf{C}}} e^{\langle \bar{z}, \xi \rangle + \langle z, \eta \rangle} \, \nu(dz) = e^{\langle \xi, \eta \rangle}, \qquad \xi, \eta \in E_{\mathbf{C}}, \tag{13}$$

$$\int_{E^*_{\mathbf{C}}} \langle z^{\otimes m}, \xi_1 \otimes \cdots \otimes \xi_m \rangle \langle \bar{z}^{\otimes n}, \eta_1 \otimes \cdots \otimes \eta_n \rangle \, \nu(dz)$$

$$= \delta_{mn} \, m! \, \langle \xi_1 \widehat{\otimes} \ldots \widehat{\otimes} \xi_m, \eta_1 \widehat{\otimes} \ldots \widehat{\otimes} \eta_m \rangle, \qquad \xi_i, \eta_j \in E_{\mathbf{C}}. \tag{14}$$

2.2 CKS-Space over Complex Gaussian Space

By modifying the Wiener–Itô–Segal isomorphism (12) we have a unitary isomorphism:

$$L^2(E^*, \mu_{1/2}) \cong \Gamma(H_{\mathbf{C}}) \tag{15}$$

uniquely specified by the correspondence:

$$\psi_\xi(x) \equiv e^{\sqrt{2}\langle x, \xi \rangle - \langle \xi, \xi \rangle/2}$$

$$\leftrightarrow \phi_\xi \equiv \left(1, \xi, \frac{\xi^{\otimes 2}}{2!}, \ldots, \frac{\xi^{\otimes n}}{n!}, \ldots \right), \quad \xi \in E_{\mathbf{C}},$$

Identifying functions on $E^*_{\mathbf{C}}$ and on $E^* \times E^*$ in a canonical manner:

$$\phi \otimes \psi(x + iy) = \phi(x)\psi(y), \qquad x, y \in E^*, \quad \phi, \psi \in L^2(E^*, \mu_{1/2}), \tag{16}$$

we have an isomorphism:

$$L^2(E^*_{\mathbf{C}}, \nu) \cong L^2(E^*, \mu_{1/2}) \otimes L^2(E^*, \mu_{1/2}). \tag{17}$$

Then, combining (15) and (17) we obtain a unitary isomorphism

$$\mathcal{I} : \Gamma(H_{\mathbf{C}}) \otimes \Gamma(H_{\mathbf{C}}) \to L^2(E^*_{\mathbf{C}}, \nu). \tag{18}$$

Moreover, applying this $\mathcal{I}$ to the Gelfand triple (10), we come to a Gelfand triple:

$$\mathcal{D} \subset L^2(E^*_{\mathbf{C}}, \nu) \subset \mathcal{D}^*, \tag{19}$$

which is referred to as the *CKS-space over the complex Gaussian space*. By construction, $\mathcal{I}$ becomes topological isomorphisms (denoted by the same symbol) from $\mathcal{W} \otimes \mathcal{W}$ onto $\mathcal{D}$ and $(\mathcal{W} \otimes \mathcal{W})^*$ onto $\mathcal{D}^*$. The canonical $\mathbf{C}$-bilinear form on $\mathcal{D}^* \times \mathcal{D}$ is denoted by $\langle\!\langle \cdot, \cdot \rangle\!\rangle_\nu$ and the norms induced by $\Gamma_\alpha(E_p)$ are denoted by $\|\cdot\|_{\nu,\pm p,\pm}$ for $p \geq 0$, i.e.,

$$\| \mathcal{I}(\phi \otimes \psi) \|_{\nu,\pm p,\pm} = \| \phi \|_{\pm p,\pm} \| \psi \|_{\pm p,\pm}, \qquad \phi, \psi \in \mathcal{W}.$$

For $\xi \in E_{\mathbf{C}}$ we define

$$\epsilon_\xi(z) = e^{\langle z, \xi \rangle}, \qquad z \in E_{\mathbf{C}}^*.$$

By a simple calculation we come to the following

Lemma 2.1 *The isomorphism $\mathcal{I}$ is uniquely determined by*

$$\mathcal{I}(\phi_\xi \otimes \phi_\eta)(z) = \epsilon_{(\xi - i\eta)/\sqrt{2}}(z)\epsilon_{(\xi + i\eta)/\sqrt{2}}(\bar{z})e^{-\langle \xi, \xi \rangle/2 - \langle \eta, \eta \rangle/2},$$

where ξ, η run over $E_{\mathbf{C}}$.

2.3   Holomorphic L^2-Functions

By means of the orthogonal relation (14) one easily obtains the following

Lemma 2.2 *For any $\phi = (f_m) \in \Gamma(H_{\mathbf{C}})$*

$$\omega_\phi(z) = \sum_{m=0}^{\infty} \langle z^{\otimes m}, f_m \rangle, \qquad z \in E_{\mathbf{C}}^*,$$

is defined in the sense of $L^2(E_{\mathbf{C}}, \nu)$ and $\phi \mapsto \omega_\phi$ is an isometric map from $\Gamma(H_{\mathbf{C}})$ into $L^2(E_{\mathbf{C}}^*, \nu)$.

We define a closed subspace of $L^2(E_{\mathbf{C}}^*, \nu)$ by

$$L^2(E_{\mathbf{C}}^*, \nu)_{\mathrm{HOL}} = \{\omega_\phi \, ; \, \phi \in \Gamma(H_{\mathbf{C}})\}.$$

Then, in view of the Wiener–Itô–Segal isomorphism $L^2(E^*, \mu) \cong \Gamma(H_{\mathbf{C}})$ we obtain a unitary isomorphism

$$L^2(E^*, \mu) \cong L^2(E_{\mathbf{C}}^*, \nu)_{\mathrm{HOL}}.$$

This is the famous Segal–Bargmann transform[8].

Lemma 2.3 *Under the identification $L^2(E_{\mathbf{C}}^*, \nu) \cong \Gamma(H_{\mathbf{C}}) \otimes \Gamma(H_{\mathbf{C}})$ given by $\mathcal{I}$ we have the following correspondence:*

$$\langle z, \zeta \rangle^m$$

$$\leftrightarrow \sum_{k=0}^{m} \binom{m}{k} \left(0, \ldots, \left(\frac{\zeta}{\sqrt{2}}\right)^{\otimes k}, 0, \ldots\right) \otimes \left(0, \ldots, \left(\frac{i\zeta}{\sqrt{2}}\right)^{\otimes(m-k)}, 0, \ldots\right),$$

$$\langle \bar{z}, \zeta \rangle^m$$

$$\leftrightarrow \sum_{k=0}^{m} \binom{m}{k} \left(0, \ldots, \left(\frac{\zeta}{\sqrt{2}}\right)^{\otimes k}, 0, \ldots\right) \otimes \left(0, \ldots, \left(\frac{\zeta}{i\sqrt{2}}\right)^{\otimes(m-k)}, 0, \ldots\right),$$

where $\zeta \in E_{\mathbf{C}}$.

PROOF. It follows from Lemma 2.1 that

$$\mathcal{I}(\phi_\xi \otimes \phi_\eta)(z) = \epsilon_{\zeta'}(z)\epsilon_\zeta(\bar{z})e^{-\langle \zeta, \zeta' \rangle},$$

where

$$\zeta = \frac{\xi + i\eta}{\sqrt{2}}, \quad \zeta' = \frac{\xi - i\eta}{\sqrt{2}}, \qquad \xi, \eta \in E_{\mathbf{C}}.$$

Then the assertion follows by Taylor expansion. ∎

Lemma 2.4 *For $\zeta \in E_{\mathbf{C}}$ and $m = 0, 1, 2, \ldots$ we put $\omega_{m,\zeta}(z) = \langle z, \zeta \rangle^m$. Then, for $p \geq 0$ we have*

$$\| \omega_{m,\zeta} \|_{\nu,p,+}^2 = m! \, | \zeta^{\otimes m} |_p^2 \, \frac{1}{2^m} \sum_{k=0}^{m} \binom{m}{k} \alpha(k)\alpha(m - k). \tag{20}$$

PROOF. For simplicity we put

$$h_k(\zeta) = (0, \ldots, \zeta^{\otimes k}, 0, \ldots), \qquad \zeta \in E_{\mathbf{C}}, \quad k \geq 0.$$

Obviously, $h_k(\zeta) \in \mathcal{W}$ and

$$\| h_k(\zeta) \|_{p,+}^2 = k!\alpha(k) \, | \zeta |_p^{2k}.$$

It follows from Lemma 2.3 that

$$\mathcal{I}^{-1}\omega_{m,\zeta} = \sum_{k=0}^{m} \binom{m}{k} h_k \left(\frac{\zeta}{\sqrt{2}} \right) \otimes h_{m-k} \left(\frac{i\zeta}{\sqrt{2}} \right).$$

Since the right hand side is an orthogonal sum with respect to the norm $\| \cdot \|_{p,+}$, we have

$$
\begin{aligned}
\| \omega_{m,\zeta} \|_{\nu,p,+}^2 &= \sum_{k=0}^{m} \binom{m}{k}^2 \left\| h_k \left(\frac{\zeta}{\sqrt{2}} \right) \otimes h_{m-k} \left(\frac{i\zeta}{\sqrt{2}} \right) \right\|_{p,+}^2 \\
&= \sum_{k=0}^{m} \binom{m}{k}^2 \left\| h_k \left(\frac{\zeta}{\sqrt{2}} \right) \right\|_{p,+}^2 \left\| h_{m-k} \left(\frac{i\zeta}{\sqrt{2}} \right) \right\|_{p,+}^2 \\
&= \sum_{k=0}^{m} \binom{m}{k}^2 k!\alpha(k) \left| \frac{\zeta}{\sqrt{2}} \right|_p^{2k} (m - k)!\alpha(m - k) \left| \frac{i\zeta}{\sqrt{2}} \right|_p^{2(m-k)} \\
&= m! | \zeta |_p^{2m} \frac{1}{2^m} \sum_{k=0}^{m} \binom{m}{k} \alpha(k)\alpha(m - k),
\end{aligned}
$$

as desired. ∎

Proposition 2.5 *For $\phi = (f_m) \in \mathcal{W}$ let $\omega_\phi \in L^2(E_{\mathbf{C}}^*, \nu)$ be defined as in Lemma 2.2. Then for any $p \geq 0$,*

$$\| \omega_\phi \|_{\nu,p,+}^2 = \sum_{m=0}^{\infty} m! \, | f_m |_p^2 \, \frac{1}{2^m} \sum_{k=0}^{m} \binom{m}{k} \alpha(k)\alpha(m-k). \qquad (21)$$

PROOF. For $f_m \in E_{\mathbf{C}}^{\widehat{\otimes} m}$ we put $\omega_m(z) = \langle z^{\otimes m}, f_m \rangle$. Then, by the polarization formula and Fourier expansion we see from Lemma 2.4 that

$$\| \omega_m \|_{\nu,p,+}^2 = m! \, | f_m |_p^2 \, \frac{1}{2^m} \sum_{k=0}^{m} \binom{m}{k} \alpha(k)\alpha(m-k), \qquad p \geq 0.$$

Since $\omega_\phi = \sum_{m=0}^{\infty} \omega_m$ is an orthogonal sum, (21) follows immediately. ∎

2.4 Holomorphic Test Functions

We define

$$\mathcal{D}_{\mathrm{HOL}} = \mathcal{D} \cap L^2(E_{\mathbf{C}}, \nu)_{\mathrm{HOL}}.$$

Let $\phi \mapsto \omega_\phi$ be the unitary map from $\Gamma(H_{\mathbf{C}})$ onto $L^2(E_{\mathbf{C}}, \nu)_{\mathrm{HOL}}$ defined in Lemma 2.2.

Proposition 2.6 *The map $\phi \mapsto \omega_\phi$ induces a topological isomorphism from $\mathcal{W}$ onto $\mathcal{D}_{\mathrm{HOL}}$.*

PROOF. Let $\phi = (f_m) \in \mathcal{W}$ and start with (21). Since $\alpha(k)\alpha(m-k) \leq C_1^m \alpha(m)$ by Lemma 1.1 (1), we have

$$
\begin{aligned}
\| \omega_\phi \|_{\nu,p,+}^2 &= \sum_{m=0}^{\infty} m! \, | f_m |_p^2 \, \frac{1}{2^m} \sum_{k=0}^{m} \binom{m}{k} \alpha(k)\alpha(m-k) \\
&\leq \sum_{m=0}^{\infty} m! \, C_1^m \alpha(m) | f_m |_p^2 \, \frac{1}{2^m} \sum_{k=0}^{m} \binom{m}{k} \\
&\leq \sum_{m=0}^{\infty} m! \, C_1^m \alpha(m) | f_m |_p^2. \qquad (22)
\end{aligned}
$$

Choose $q \geq 0$ in such a way that $C_1 \rho^{2q} \leq 1$, where $\rho = \|A^{-1}\|_{\mathrm{OP}} = 1/2$. Then, (22) becomes

$$\| \omega_\phi \|_{\nu,p,+}^2 \leq \sum_{m=0}^{\infty} m! \, C_1^m \alpha(m) \rho^{2qm} | f_m |_{p+q}^2 \leq \| \phi \|_{p+q,+}^2, \qquad \phi \in \mathcal{W}. \qquad (23)$$

A similar argument using Lemma 1.1 (2) yields

$$\|\phi\|_{p,+} \leq \|\omega_\phi\|_{\nu,p+r,+}, \qquad \phi \in \mathcal{W}, \tag{24}$$

where $r \geq 0$ is chosen in such a way that $C_2\rho^{2q} \leq 1$. The assertion is a consequence of two norm estimates (23) and (24). $\blacksquare$

3 Representation of White Noise Functions

3.1 S-transform and Characterization Theorem

The S-transform of $\Phi \in \mathcal{W}^*$ is defined by

$$S\Phi(\xi) = \langle\!\langle \Phi, \phi_\xi \rangle\!\rangle, \qquad \xi \in E_{\mathbf{C}}.$$

It follows from Proposition 1.3 that the S-transform determines a white noise function uniquely. In fact, for $\Phi = (F_n)$ we have

$$S\Phi(\xi) = \sum_{n=0}^{\infty} \langle F_n, \xi^{\otimes n} \rangle, \qquad \xi \in E_{\mathbf{C}}. \tag{25}$$

Moreover, we have the following fundamental result known as the characterization theorem for S-transform[6,25].

Theorem 3.1 *A **C**-valued function F defined on $E_{\mathbf{C}}$ is the S-transform of a white noise distribution $\Phi \in \mathcal{W}^*$ if and only if*

(F1) *for any $\xi, \xi_1 \in E_{\mathbf{C}}$, $z \mapsto F(z\xi + \xi_1)$ is entire holomorphic on **C**;*

(F2) *there exist some $C \geq 0$ and $p \geq 0$ such that*

$$|F(\xi)|^2 \leq CG_\alpha(|\xi|_p^2), \qquad \xi \in E_{\mathbf{C}}.$$

In that case,

$$\|\Phi\|_{-(p+q),-}^2 \leq C\widetilde{G}_\alpha(\|A^{-q}\|_{\mathrm{HS}}^2),$$

for any $q > 1/2$ such that $\widetilde{G}_\alpha(\|A^{-q}\|_{\mathrm{HS}}^2) < \infty$. (This choice is always possible since $\|A^{-q}\|_{\mathrm{HS}} \to 0$ as $q \to \infty$.)

3.2 Holomorphic Distributions

For $\phi \in \mathcal{W}$ the S-transform $S\phi$ is naturally extended to a function defined on $E_{\mathbf{C}}^*$ by

$$S\phi(z) = \langle\!\langle \phi_z, \phi \rangle\!\rangle, \qquad z \in E_{\mathbf{C}}^*,$$

where

$$\phi_z = \left(1, z, \frac{z^{\otimes 2}}{2!}, \ldots, \frac{z^{\otimes n}}{n!}, \ldots \right) \in \mathcal{W}^*$$

and is also referred to as the exponential vector. Then, for $\phi = (f_n) \in \mathcal{W}$ we have

$$S\phi(z) = \sum_{n=0}^{\infty} \langle z^{\otimes n}, f_n \rangle = \omega_\phi(z), \qquad z \in E_{\mathbf{C}}^*.$$

by Lemma 2.2. In other words, the S-transform, which is defined for all $\Phi \in \mathcal{W}^*$ and $S\Phi$ is an entire function on $E_{\mathbf{C}}$, and the Segal–Bargmann transform, which is defined for all $\phi \in \Gamma(H_{\mathbf{C}})$ and ω_ϕ is an entire function on $E_{\mathbf{C}}^*$, coincide on a common domain.

Moreover, we shall prove that a formal notation

$$\omega_\Phi(z) = \sum_{n=0}^{\infty} \langle z^{\otimes n}, F_n \rangle, \qquad \Phi = (F_n) \in \mathcal{W}^*, \tag{26}$$

is given a meaning as an element of $\mathcal{D}^*$. (Recall that (26) converges if $\Phi \in \mathcal{W}$ or if $\Phi \in \Gamma(H_{\mathbf{C}})$.) By a similar argument as in the proof of Proposition 2.5 we come to

Proposition 3.2 *For $\phi = (f_m) \in \mathcal{W}$ let ω_ϕ be defined as in (26). Then for any $p \geq 0$,*

$$\| \omega_\phi \|_{\nu, -p, -}^2 = \sum_{m=0}^{\infty} m! \, | f_m |_{-p}^2 \, \frac{1}{2^m} \sum_{k=0}^{m} \binom{m}{k} \frac{1}{\alpha(k)\alpha(m-k)}. \tag{27}$$

Then, obviously the map $\phi \mapsto \omega_\phi$ extends an isomorphism from $\mathcal{W}^*$ into $\mathcal{D}^*$ and we adopt a formal notation for ω_Φ as in (26). We put

$$\mathcal{D}_{\mathrm{HOL}}^* = \{ \omega_\Phi \, ; \, \Phi \in \mathcal{W}^* \}.$$

Then we have already seen that $\Phi \mapsto \omega_\Phi$ is a topological isomorphism from $\mathcal{W}^*$ onto $\mathcal{D}_{\mathrm{HOL}}^*$. In the sense that the map $\Phi \mapsto \omega_\Phi$ is an extension of the S-transform $S\Phi$ we may state

Theorem 3.3 *The S-transform extends the Segal–Bargmann transform and is a topological isomorphism from $\mathcal{W}^*$ onto $\mathcal{D}_{\mathrm{HOL}}^*$.*

A holomorphic distribution $\Omega \in \mathcal{D}^*_{\mathrm{HOL}}$ feels only anti-holomorphic test functions. Define the space of anti-holomorphic L^2-functions by

$$L^2(E^*_{\mathbf{C}}, \nu)_{\mathrm{AH}} = \{\phi \in L^2(E^*_{\mathbf{C}}, \nu)\,;\, \bar{\phi} \in L^2(E^*_{\mathbf{C}}, \nu)_{\mathrm{HOL}}\}$$

and the space of anti-holomorphic test functions by

$$\mathcal{D}_{\mathrm{AH}} = \mathcal{D} \cap L^2(E^*_{\mathbf{C}}, \nu)_{\mathrm{AH}}.$$

Then for $\Omega \in \mathcal{D}^*_{\mathrm{HOL}}$ and $\omega \in \mathcal{D}_{\mathrm{AH}}$ we have

$$\langle\!\langle \Omega,\, \omega \rangle\!\rangle_\nu = \sum_{n=0}^{\infty} n!\, \langle F_n,\, f_n \rangle, \tag{28}$$

where

$$\Omega(z) = \sum_{n=0}^{\infty} \langle z^{\otimes n},\, F_n \rangle, \qquad \omega(z) = \sum_{n=0}^{\infty} \langle \bar{z}^{\otimes n},\, f_n \rangle.$$

3.3 Coherent State Representation and Inverse S-transform

By definition, for $\xi \in E_{\mathbf{C}}$ we have

$$\epsilon_\xi(z) = e^{\langle z,\, \xi \rangle} = \psi_{\xi/\sqrt{2}}(x)\psi_{i\xi/\sqrt{2}}(y), \qquad z = x + iy \in E^*_{\mathbf{C}}.$$

In other words,

$$\mathcal{I}(\phi_{\xi/\sqrt{2}} \otimes \phi_{i\xi/\sqrt{2}}) = \epsilon_\xi \tag{29}$$

and, in particular, $\epsilon_\xi \in \mathcal{D}$.

Lemma 3.4 *For any $\rho \in \mathcal{D}^*$ there exists a unique $\Phi \in \mathcal{W}^*$ such that*

$$S\Phi(\xi) = \langle\!\langle \rho,\, \epsilon_\xi \rangle\!\rangle_\nu, \qquad \xi \in E_{\mathbf{C}}. \tag{30}$$

PROOF. Denote by $F(\xi)$ the right hand side of (30). It is obvious that F satisfies condition (F1) in Theorem 3.1. We shall prove (F2). Choose $p \geq 0$ such that $\|\rho\|_{\nu,-p,-} < \infty$. Then in view of (29) we see that

$$\begin{aligned}
|F(\xi)|^2 &\leq \|\rho\|^2_{\nu,-p,-} \|\epsilon_\xi\|^2_{\nu,p,+} \\
&= \|\rho\|^2_{\nu,-p,-} \|\phi_{\xi/\sqrt{2}}\|^2_{p,+} \|\phi_{i\xi/\sqrt{2}}\|^2_{p,+} \\
&= \|\rho\|^2_{\nu,-p,-}\, G_\alpha\!\left(\left|\frac{\xi}{\sqrt{2}}\right|^2_p\right) G_\alpha\!\left(\left|\frac{i\xi}{\sqrt{2}}\right|^2_p\right).
\end{aligned} \tag{31}$$

In view of Proposition 1.2 (2) we have

$$G_\alpha\!\left(\left|\frac{\xi}{\sqrt{2}}\right|^2_p\right) G_\alpha\!\left(\left|\frac{i\xi}{\sqrt{2}}\right|^2_p\right) \leq G_\alpha(C_1\, |\xi|^2_p) \leq G_\alpha(C_1\rho^{2q}\, |\xi|^2_{p+q}),$$

where $\rho = \|A^{-1}\|_{OP} = 1/2$. Choose $q \geq 0$ such that $C_1 \rho^{2q} \leq 1$. Then we come to

$$|F(\xi)|^2 \leq \|\rho\|_{\nu,-p,-}^2 \, G_\alpha(|\xi|_{p+q}^2),$$

and condition (F2) is fulfilled. Thus, by Thoerem 3.1, F is the S-transform of a white noise distribution in $\mathcal{W}^*$. ∎

The white noise distribution Φ defined as in (30) is denoted by

$$\Phi = \int_{E_{\mathbf{C}}^*} \rho(z)\phi_z \nu(dz)$$

and is called a *coherent state representation*.

Theorem 3.5 *For any $\Phi \in \mathcal{W}^*$, the S-transform $S\Phi$ being regarded as a holomorphic distribution on the complex Gaussian space, it holds that*

$$\Phi = \int_{E_{\mathbf{C}}^*} S\Phi(\bar{z})\phi_z \nu(dz).$$

In particular, every white noise distribution Φ admits a coherent state representation.

PROOF. It follows from Theorem 3.3 that

$$\rho(z) \equiv S\Phi(\bar{z}) = \sum_{n=0}^{\infty} \langle \bar{z}^{\otimes n}, F_n \rangle$$

belongs to $\mathcal{D}^*$ and by Lemma 3.4 there exists a unique $\Psi \in \mathcal{W}^*$ satisfying $S\Psi(\xi) = \langle\!\langle \rho, \epsilon_\xi \rangle\!\rangle_\nu$ for $\xi \in E_{\mathbf{C}}$. Note that

$$\epsilon_\xi(z) = e^{\langle z, \xi \rangle} = \sum_{n=0}^{\infty} \left\langle z^{\otimes n}, \frac{\xi^{\otimes n}}{n!} \right\rangle.$$

Hence by (28),

$$S\Psi(\xi) = \langle\!\langle \rho, \epsilon_\xi \rangle\!\rangle_\nu = \sum_{n=0}^{\infty} n! \left\langle F_n, \frac{\xi^{\otimes n}}{n!} \right\rangle = \sum_{n=0}^{\infty} \langle F_n, \xi^{\otimes n} \rangle,$$

which coincides with $S\Phi(\xi)$. Therefore $\Phi = \Psi$. ∎

A relavant formula was obtained by Berezansky–Kondratiev[3] by means of duality argument over complex Gaussian space in a different realization. The inverse S-transform was also discussed within the real Gaussian integral[15,17].

4 Representation of White Noise Operators

4.1 White Noise Operators, Symbols and Kernels

A continuous operator from $\mathcal{W}$ into $\mathcal{W}^*$ is in general referred to as a *white noise operator*. Let $\mathcal{L}(\mathcal{W}, \mathcal{W}^*)$ denote the space of white noise operators. The *symbol* of $\Xi \in \mathcal{L}(\mathcal{W}, \mathcal{W}^*)$ is by definition a **C**-valued function on $E_{\mathbf{C}} \times E_{\mathbf{C}}$ defined by

$$\widehat{\Xi}(\xi, \eta) = \langle\!\langle \Xi\phi_\xi, \phi_\eta \rangle\!\rangle, \qquad \xi, \eta \in E_{\mathbf{C}}.$$

The symbol uniquely specifies a white noise operator by Proposition 1.3. Moreover, by the kernel theorem there exists a unique $\Xi^K \in (\mathcal{W} \otimes \mathcal{W})^*$ such that

$$\widehat{\Xi}(\xi, \eta) = \langle\!\langle \Xi\phi_\xi, \phi_\eta \rangle\!\rangle = \langle\!\langle \Xi^K, \phi_\xi \otimes \phi_\eta \rangle\!\rangle, \qquad \xi, \eta \in E_{\mathbf{C}}.$$

This Ξ^K is called the *kernel* of Ξ.

We now state the characterization for operator symbols[4,18].

Theorem 4.1 *A function* $\Theta : E_{\mathbf{C}} \times E_{\mathbf{C}} \to \mathbf{C}$ *is the symbol of an operator* $\Xi \in \mathcal{L}(\mathcal{W}, \mathcal{W}^*)$ *if and only if*

(O1) *for any* $\xi, \xi_1, \eta, \eta_1 \in E_{\mathbf{C}}$, $(z, w) \mapsto \Theta(z\xi + \xi_1, w\eta + \eta_1)$ *is entire holomorphic on* $\mathbf{C} \times \mathbf{C}$;

(O2) *there exist constant numbers* $C \geq 0$ *and* $p \geq 0$ *such that*

$$|\Theta(\xi, \eta)|^2 \leq C G_\alpha(|\xi|_p^2) G_\alpha(|\eta|_p^2), \qquad \xi, \eta \in E_{\mathbf{C}}.$$

In that case

$$\|\Xi\phi\|^2_{-(p+q),-} \leq C\widetilde{G}^2_\alpha(\|A^{-q}\|^2_{\mathrm{HS}})\|\phi\|^2_{p+q,+}, \qquad \phi \in \mathcal{W}, \tag{32}$$

where $q > 1/2$ *is taken as* $\widetilde{G}_\alpha(\|A^{-q}\|^2_{\mathrm{HS}}) < \infty$.

4.2 Diagonal Coherent State Representation

With each $z \in E_{\mathbf{C}}^*$ we associate $Q_z \in \mathcal{L}(\mathcal{W}, \mathcal{W}^*)$ by the formula:

$$Q_z\phi = \langle\!\langle \phi_{\bar{z}}, \phi \rangle\!\rangle \phi_z, \qquad \phi \in \mathcal{W}.$$

Note here that both maps $z \mapsto \phi_z \in \mathcal{W}^*$ and $z \mapsto Q_z \in \mathcal{L}(\mathcal{W}, \mathcal{W}^*)$ are continuous. The symbol of Q_z is given by

$$\widehat{Q}_z(\xi, \eta) \equiv q_{\xi,\eta}(z) = e^{\langle \bar{z}, \xi \rangle + \langle z, \eta \rangle}, \qquad z \in E_{\mathbf{C}}^*, \quad \xi, \eta \in E_{\mathbf{C}}. \tag{33}$$

Then, for $z = x + iy$ we obtain

$$q_{\xi,\eta}(x + iy) = e^{\langle x,\, \xi+\eta\rangle} e^{\langle y,\, i(-\xi+\eta)\rangle}$$
$$= e^{\langle \xi,\, \eta\rangle} \psi_{(\xi+\eta)/\sqrt{2}}(x) \psi_{i(-\xi+\eta)/\sqrt{2}}(y),$$

and hence

$$q_{\xi,\eta} = e^{\langle \xi,\, \eta\rangle} \mathcal{I}(\phi_{(\xi+\eta)/\sqrt{2}} \otimes \phi_{i(-\xi+\eta)/\sqrt{2}}). \tag{34}$$

In particular, $q_{\xi,\eta} \in \mathcal{D}$.

Lemma 4.2 *For $\rho \in \mathcal{D}^*$ there is a unique operator $\Xi \in \mathcal{L}(\mathcal{W}, \mathcal{W}^*)$ such that*

$$\langle\!\langle \Xi \phi_\xi,\, \phi_\eta \rangle\!\rangle = \langle\!\langle \rho,\, q_{\xi,\eta} \rangle\!\rangle_\nu, \qquad \xi, \eta \in E_{\mathbf{C}}. \tag{35}$$

The proof is similar to that of Lemma 3.4 with the help of Theorem 4.1. The operator Ξ defined by (35) is denoted by

$$\Xi = \int_{E_{\mathbf{C}}^*} \rho(z) Q_z\, \nu(dz) \tag{36}$$

and is called the *diagonal coherent state representation*. The next result was proved first by Obata[20] and we give a different proof here.

Theorem 4.3 *Every white noise operator $\Xi \in \mathcal{L}(\mathcal{W}, \mathcal{W}^*)$ admits a unique diagonal coherent state representation.*

PROOF. We shall first define an operator $\mathcal{G} \in \mathcal{L}(\mathcal{W} \otimes \mathcal{W}, \mathcal{W} \otimes \mathcal{W})$ uniquely specified by

$$\mathcal{G}(\phi_\xi \otimes \phi_\eta) = e^{-\langle \xi,\, \xi\rangle/2 - \langle \eta,\, \eta\rangle/2} \phi_{(\xi+i\eta)/\sqrt{2}} \otimes \phi_{(\xi-i\eta)/\sqrt{2}}, \qquad \xi, \eta \in E_{\mathbf{C}}.$$

Since the exponential vectors are linearly independent, so are $\{\phi_\xi \otimes \phi_\eta;\ \xi, \eta \in E_{\mathbf{C}}\}$ and $\mathcal{G}$ is well defined on a subspace spaned by such tensor products of exponential vectors. It is then sufficient to check that the function

$$\Theta(\xi, \eta; \xi', \eta') = \langle\!\langle \mathcal{G}(\phi_\xi \otimes \phi_\eta),\, \phi_{\xi'} \otimes \phi_{\eta'} \rangle\!\rangle$$
$$= e^{-\langle \xi,\, \xi\rangle/2 - \langle \eta,\, \eta\rangle/2} \langle\!\langle \phi_{(\xi+i\eta)/\sqrt{2}},\, \phi_{\xi'} \rangle\!\rangle \langle\!\langle \phi_{(\xi-i\eta)/\sqrt{2}},\, \phi_{\eta'} \rangle\!\rangle \tag{37}$$

satisfies the conditions in the characterization theorem in the case of multi-variables (see Theorem A.6). In fact, condition (O1) is rather obvious. To see (O2) we take an arbitrary $p \geq 0$. We see easily that

$$|\langle\!\langle \phi_{(\xi+i\eta)/\sqrt{2}},\, \phi_{\xi'} \rangle\!\rangle|^2 \leq \| \phi_{(\xi+i\eta)/\sqrt{2}} \|_{p,+}^2 \| \phi_{\xi'} \|_{-p,-}^2$$
$$= G_\alpha\left(\left| \frac{\xi + i\eta}{\sqrt{2}} \right|_p^2 \right) G_{1/\alpha}(|\xi'|_{-p}^2)$$
$$\leq G_\alpha(|\xi|_p^2 + |\eta|_p^2) G_{1/\alpha}(|\xi'|_{-p}^2).$$

Then by Proposition 1.2 (3) we obtain

$$|\langle\!\langle \phi_{(\xi+i\eta)/\sqrt{2}}, \phi_{\xi'} \rangle\!\rangle|^2 \leq G_\alpha(C_2\,|\xi\,|_p^2)G_\alpha(C_2\,|\eta\,|_p^2)G_{1/\alpha}(|\xi'\,|_{-p}^2). \tag{38}$$

Similarly,

$$|\langle\!\langle \phi_{(\xi-i\eta)/\sqrt{2}}, \phi_{\eta'} \rangle\!\rangle|^2 \leq G_\alpha(C_2\,|\xi\,|_p^2)G_\alpha(C_2\,|\eta\,|_p^2)G_{1/\alpha}(|\eta'\,|_{-p}^2). \tag{39}$$

On the other hand, it follows from Proposition 1.2 (4) that

$$|e^{-\langle\xi,\xi\rangle/2}|^2 \leq e^{|\langle\xi,\xi\rangle|} \leq G_\alpha(C_3\rho^{2p}\,|\xi\,|_p^2), \tag{40}$$

where $\rho = \|A^{-1}\|_{\mathrm{OP}} = 1/2$, and similarly,

$$|e^{-\langle\eta,\eta\rangle/2}|^2 \leq G_\alpha(C_3\rho^{2p}\,|\eta\,|_p^2). \tag{41}$$

Combining (38)–(41), we obtain an estimate for (37) as follows:

$$
\begin{aligned}
|\Theta(\xi,\eta;\xi',\eta')|^2 \leq\ & G_\alpha(C_2\,|\xi\,|_p^2)G_\alpha(C_2\,|\eta\,|_p^2)G_{1/\alpha}(|\xi'\,|_{-p}^2) \\
& \times G_\alpha(C_2\,|\xi\,|_p^2)G_\alpha(C_2\,|\eta\,|_p^2)G_{1/\alpha}(|\eta'\,|_{-p}^2) \\
& \times G_\alpha(C_3\rho^{2p}\,|\xi\,|_p^2)G_\alpha(C_3\rho^{2p}\,|\eta\,|_p^2)
\end{aligned}
\tag{42}
$$

By virtue of Proposition 1.2 (2) there exists some $C > 0$ such that

$$G_\alpha(C_2\,|\xi\,|_p^2)G_\alpha(C_2\,|\xi\,|_p^2)G_\alpha(C_3\rho^{2p}\,|\xi\,|_p^2) \leq G_\alpha(C\,|\xi\,|_p^2).$$

Now choose $q \geq 0$ satisfying $C\rho^{2q} \leq 1$ to have

$$G_\alpha(C_2\,|\xi\,|_p^2)G_\alpha(C_2\,|\xi\,|_p^2)G_\alpha(C_3\rho^{2p}\,|\xi\,|_p^2) \leq G_\alpha(|\xi\,|_{p+q}^2).$$

Together with a similar inequality for η we finally conclude that (42) becomes

$$|\Theta(\xi,\eta;\xi',\eta')|^2 \leq G_\alpha(|\xi\,|_{p+q}^2)G_\alpha(|\eta\,|_{p+q}^2)G_{1/\alpha}(|\xi'\,|_{-p}^2)G_{1/\alpha}(|\eta'\,|_{-p}^2),$$

which shows the desired condition (O2). Thus $\mathcal{G} \in \mathcal{L}(\mathcal{W} \otimes \mathcal{W}, \mathcal{W} \otimes \mathcal{W})$.

Similarly, $\mathcal{H} \in \mathcal{L}(\mathcal{W} \otimes \mathcal{W}, \mathcal{W} \otimes \mathcal{W})$ is uniquely specified by

$$\mathcal{H}(\phi_\xi \otimes \phi_\eta) = e^{\langle\xi,\eta\rangle}\phi_{(\xi+\eta)/\sqrt{2}} \otimes \phi_{i(-\xi+\eta)/\sqrt{2}}, \qquad \xi,\eta \in E_{\mathbf{C}}.$$

Moreover,

$$\mathcal{G}\mathcal{H}(\phi_\xi \otimes \phi_\eta) = \mathcal{H}\mathcal{G}(\phi_\xi \otimes \phi_\eta) = \phi_\xi \otimes \phi_\eta, \qquad \xi,\eta \in E_{\mathbf{C}}.$$

Therefore, $\mathcal{G}$ is an automorphism of $\mathcal{W} \otimes \mathcal{W}$ and $\mathcal{G}^{-1} = \mathcal{H}$. Now, suppose we are given an arbitrary operator $\Xi \in \mathcal{L}(\mathcal{W}, \mathcal{W}^*)$. Then

$$
\begin{aligned}
\langle\!\langle \Xi\phi_\xi, \phi_\eta \rangle\!\rangle &= \langle\!\langle \Xi^K, \phi_\xi \otimes \phi_\eta \rangle\!\rangle \\
&= \langle\!\langle \Xi^K, \mathcal{G}\mathcal{H}(\phi_\xi \otimes \phi_\eta) \rangle\!\rangle \\
&= \langle\!\langle \mathcal{G}^*\Xi^K, \mathcal{H}(\phi_\xi \otimes \phi_\eta) \rangle\!\rangle \\
&= \langle\!\langle \mathcal{G}^*\Xi^K, \phi_{(\xi+\eta)/\sqrt{2}} \otimes \phi_{i(-\xi+\eta)/\sqrt{2}} \rangle\!\rangle e^{\langle\xi,\eta\rangle}.
\end{aligned}
$$

Using (34) we see immediately that

$$\langle\!\langle \Xi\phi_\xi, \phi_\eta \rangle\!\rangle = \langle\!\langle \mathcal{I}\mathcal{G}^*\Xi^K, q_{\xi,\eta} \rangle\!\rangle_\nu.$$

This means that Ξ admits a diagonal coherent state representation:

$$\Xi = \int_{E_{\mathbf{C}}^*} \rho(z) Q_z \, \nu(dz), \qquad \rho = \mathcal{I}\mathcal{G}^*\Xi^K \in \mathcal{D}^*.$$

The uniqueness of ρ is obvious since $\mathcal{I}\mathcal{G}^*$ is an isomorphism from $(\mathcal{W} \otimes \mathcal{W})^*$ onto $\mathcal{D}^*$. ∎

4.3 Examples

Proposition 4.4 (Resolution of the identity) *It holds that*

$$I = \int_{E_{\mathbf{C}}^*} Q_z \, \nu(dz).$$

For the proof we need only to compute the symbols of both sides. The prototype of the above formula was given by Klauder[12], see also Klauder–Skagerstam[13] for various developments. From Proposition 4.4 a unitarity criterion is derived[22] and is applied to quantum stochastic differential equations involving square of white noise[23].

With each $\kappa_{l,m} \in (E_{\mathbf{C}}^{\otimes(l+m)})^*$ we associate a white noise operator by a formal integral expression:

$$\Xi_{l,m}(\kappa_{l,m}) = \int_{\mathbf{R}^{l+m}} \kappa_{l,m}(s_1,\ldots,s_l,t_1,\ldots,t_m)$$
$$a_{s_1}^* \ldots a_{s_l}^* a_{t_1} \ldots a_{t_m} ds_1 \ldots ds_l dt_1 \ldots dt_m. \qquad (43)$$

This is called an *integral kernel operator*. It is known[4,18] that for any $\kappa_{l,m} \in (E_{\mathbf{C}}^{\otimes(l+m)})^*$, the integral kernel operator $\Xi_{l,m}(\kappa_{l,m})$ always belongs to $\mathcal{L}(\mathcal{W}, \mathcal{W}^*)$. The symbol is easily obtained:

$$\widehat{\Xi_{l,m}(\kappa)}(\xi,\eta) = \langle \kappa, \eta^{\otimes l} \otimes \xi^{\otimes m} \rangle \, e^{\langle \xi, \eta \rangle}.$$

Moreover, every $\Xi \in \mathcal{L}(\mathcal{W}, \mathcal{W}^*)$ admits an infinite series expansion:

$$\Xi = \sum_{l,m=0}^{\infty} \Xi_{l,m}(\kappa_{l,m}), \qquad (44)$$

where the right hand side converges in $\mathcal{L}(\mathcal{W}, \mathcal{W}^*)$.

Proposition 4.5 *For $F \in (E_{\mathbf{C}}^{\otimes m})^*$ we have*

$$\Xi_{0,m}(F) = \int_{E_{\mathbf{C}}^*} \langle z^{\otimes m}, F \rangle \, Q_z \, \nu(dz), \qquad (45)$$

$$\Xi_{m,0}(F) = \int_{E_{\mathbf{C}}^*} \langle \bar{z}^{\otimes m}, F \rangle \, Q_z \, \nu(dz). \qquad (46)$$

PROOF. Put $\rho(z) = \langle z^{\otimes m}, F \rangle$. As is easily verified by definition, $\rho \in \mathcal{D}^*$ and a white noise operator

$$\Xi = \int_{E_{\mathbf{C}}^*} \langle z^{\otimes m}, F \rangle \, Q_z \, \nu(dz)$$

is defined. Then the operator symbol is given by

$$\widehat{\Xi}(\xi, \eta) = \langle\!\langle \rho, q_{\xi,\eta} \rangle\!\rangle = \int_{E_{\mathbf{C}}^*} \langle z^{\otimes m}, F \rangle \, e^{\langle \bar{z}, \xi \rangle + \langle z, \eta \rangle} \, \nu(dz).$$

Then using the orthogonal relation (14), we obtain with no difficulty

$$\widehat{\Xi}(\xi, \eta) = \langle F, \xi^{\otimes m} \rangle \, e^{\langle \xi, \eta \rangle} = \Xi_{0,m}(F)\widehat{}(\xi, \eta),$$

which proves (45). By duality (46) is obtained immediately. $\blacksquare$

For $F = \delta_t \in E^*$ we write naturally $\langle z, F \rangle = \langle z, \delta_t \rangle = z(t)$. Then $\{z(t)\}$ is nothing but the complex white noise[9]. As a special case of Proposition 4.5, we have the following

Proposition 4.6 (Quantum white noise)

$$a_t = \int_{E_{\mathbf{C}}^*} z(t) \, Q_z \nu(dz), \qquad a_t^* = \int_{E_{\mathbf{C}}^*} \overline{z(t)} \, Q_z \nu(dz).$$

5 Wick Product and Overcompleteness of Exponential Vectors

5.1 Wick Product of White Noise Functions

For $\Phi_1, \Phi_2 \in \mathcal{W}^*$ there exists a unique $\Psi \in \mathcal{W}^*$ such that

$$S\Psi(\xi) = S\Phi_1(\xi) \cdot S\Phi_2(\xi), \qquad \xi \in E_{\mathbf{C}}.$$

The verification is simple with the help of Theorem 3.1. In that case we write $\Psi = \Phi_1 \diamond \Phi_2$.

Lemma 5.1 *For $\Phi \in \mathcal{W}^*$ fixed, the map $W_\Phi : \phi \mapsto \Phi \diamond \phi$ is a continuous linear operator from $\mathcal{W}$ into $\mathcal{W}^*$.*

168

PROOF. We compute the symbol.

$$\langle\!\langle W_\Phi \phi_\xi, \phi_\eta \rangle\!\rangle = \langle\!\langle \Phi \diamond \phi_\xi, \phi_\eta \rangle\!\rangle = S(\Phi \diamond \phi_\xi)(\eta)$$
$$= S\Phi(\eta) \cdot S\phi_\xi(\eta) = \langle\!\langle \Phi, \phi_\eta \rangle\!\rangle \, e^{\langle \xi, \eta \rangle}. \tag{47}$$

Choosing $p \geq 0$ such that $\|\Phi\|_{-p,-} < \infty$, we come to

$$|\langle\!\langle W_\Phi \phi_\xi, \phi_\eta \rangle\!\rangle|^2 \leq \|\Phi\|_{-p,-}^2 \, \|\phi_\eta\|_{p,+}^2 \, e^{2|\langle \xi, \eta \rangle|}$$
$$\leq \|\Phi\|_{-p,-}^2 \, G_\alpha(|\eta|_p^2) \, e^{|\xi|_p^2 + |\eta|_p^2}, \tag{48}$$

where we used a simple inequality: $2|\langle \xi, \eta \rangle| \leq |\xi|_0^2 + |\eta|_0^2 \leq |\xi|_p^2 + |\eta|_p^2$.
Now, with the help of Proposition 1.2 we see that (48) becomes

$$|\langle\!\langle W_\Phi \phi_\xi, \phi_\eta \rangle\!\rangle|^2 \leq \|\Phi\|_{-p,-}^2 \, G_\alpha(2|\eta|_p^2) G_\alpha(|\xi|_p^2)$$
$$\leq \|\Phi\|_{-p,-}^2 \, G_\alpha(|\eta|_{p+q}^2) G_\alpha(|\xi|_{p+q}^2),$$

where $q \geq 0$ is taken in such a way that $2|\eta|_p^2 \leq 2\rho^{2q}|\eta|_{p+q}^2 \leq |\eta|_{p+q}^2$. Then
the assertion follows from the characterization theorem for operator symbols
(Theorem 4.1). ∎

The operator $W_\Phi \in \mathcal{L}(\mathcal{W}, \mathcal{W}^*)$ defined in Lemma 5.1 is called the *Wick
multiplication operator* associated with $\Phi \in \mathcal{W}^*$.

Theorem 5.2 *Let W_Φ be the Wick multiplication operator by $\Phi \in \mathcal{W}^*$. Then*

$$W_\Phi = \int_{E_{\mathbf{C}}^*} S\Phi(\bar{z}) Q_z \nu(dz). \tag{49}$$

PROOF. Let $\Phi = (F_m)$. Then

$$S\Phi(\bar{z}) = \sum_{m=0}^{\infty} \langle \bar{z}^{\otimes m}, F_m \rangle$$

as an element of $\mathcal{D}^*$, see also the proof of Theorem 3.5. Consider the diagonal
coherent state representation:

$$\Xi = \int_{E_{\mathbf{C}}^*} S\Phi(\bar{z}) Q_z \, \nu(dz).$$

It then follows from Proposition 4.5 that

$$\Xi = \sum_{m=0}^{\infty} \Xi_{m,0}(F_m),$$

and therefore,

$$\widehat{\Xi}(\xi, \eta) = \sum_{m=0}^{\infty} \langle F_m, \eta^{\otimes m} \rangle \, e^{\langle \xi, \eta \rangle} = S\Phi(\eta) e^{\langle \xi, \eta \rangle} = \langle\!\langle \Phi, \phi_\eta \rangle\!\rangle \, e^{\langle \xi, \eta \rangle}.$$

This coincides with (47) and hence $\Xi = W_\Phi$ as desired. $\blacksquare$

Thus the diagonal coherent state representation of a Wick multiplication operator is directly related to the S-transform. As an immediate application, the inversion formula for the S-transform is obtained. Since $W_\Phi \phi_0 = \Phi$ and $Q_z \phi_0 = \phi_z$, it is sufficient to apply (49) to ϕ_0.

5.2   Overcompleteness of Exponential Vectors

Let $\mathcal{D}_{\mathrm{AH}}^{\perp}$ denote the space of all $\rho \in \mathcal{D}^*$ which annihilate $\mathcal{D}_{\mathrm{AH}}$, that is, all $\rho \in \mathcal{D}^*$ such that $\langle\!\langle \rho, \phi \rangle\!\rangle = 0$ for all $\phi \in \mathcal{D}_{\mathrm{HOL}} \equiv \mathcal{D} \cap L^2(E_{\mathbf{C}}^*, \nu)_{\mathrm{HOL}}$. Recall that $\langle\!\langle \cdot, \cdot \rangle\!\rangle$ is a $\mathbf{C}$-bilinear form by our convention.

Theorem 5.3 *For $\rho \in \mathcal{D}^*$ put*

$$\Phi = \int_{E_{\mathbf{C}}^*} \rho(z) \phi_z \, \nu(dz), \qquad \Xi = \int_{E_{\mathbf{C}}^*} \rho(z) Q_z \, \nu(dz).$$

Then the following four conditions are equivalent:

(i) $\rho \in \mathcal{D}_{\mathrm{AH}}^{\perp}$;

(ii) $\Phi = 0$;

(iii) Ξ *annihilates the vacuum:* $\Xi \phi_0 = 0$;

(iv) Ξ *is a quantum stochastic integral against the annihilation process, i.e., there exists* $L \in E_{\mathbf{C}}^* \otimes \mathcal{L}(\mathcal{W}, \mathcal{W}^*)$ *such that*

$$\Xi = \int_{\mathbf{R}} L(t) a_t \, dt.$$

PROOF.   (i) $\Longrightarrow$ (ii). By Lemma 3.4,

$$S\Phi(\xi) = \langle\!\langle \rho, \epsilon_\xi \rangle\!\rangle, \qquad \xi \in E_{\mathbf{C}}.$$

Since $\epsilon_\xi \in \mathcal{D}_{\mathrm{HOL}}$, the assertion follows immediately.

(ii) $\Longrightarrow$ (i). It is easily verified that $t \mapsto \langle\!\langle \rho, \epsilon_{t\xi} \rangle\!\rangle$ is entire holomorphic on $\mathbf{C}$. Then from $0 = S\Phi(t\xi) = \langle\!\langle \rho, \epsilon_{t\xi} \rangle\!\rangle$ for all $z \in \mathbf{C}$ it follows that

$$\langle\!\langle \rho, \omega_m \rangle\!\rangle = 0, \qquad \omega_m(z) = \langle z^{\otimes m}, \xi^{\otimes m} \rangle.$$

This being valid for all $\xi \in E_{\mathbf{C}}$, we conclude that $\rho \in \mathcal{D}_{\mathrm{AH}}^{\perp}$.

(ii) $\iff$ (iii) is obvious from $Q_z\phi_0 = \phi_z$.

(iii) $\iff$ (iv). Let $\Xi = \sum_{l,m=0}^{\infty} \Xi_{l,m}(\kappa_{l,m})$ be the expansion of the form (44). Then condition (iii) is equivalent to $\langle\!\langle \Xi\phi_0, \phi_\eta \rangle\!\rangle = \widehat{\Xi}(0,\eta) = 0$ for all $\eta \in E_{\mathbf{C}}$, that is,

$$\widehat{\Xi}(0,\eta) = \sum_{l=0}^{\infty} \langle \kappa_{l,0}, \eta^{\otimes l} \rangle = 0, \qquad \eta \in E_{\mathbf{C}}.$$

This is equivalent to $\kappa_{l,0} = 0$ for all $l \geq 0$. Therefore Ξ is a sum of integral kernel operators involving one or more annihilation operators. We then see that such an operator is a quantum stochastic integral (in a broad sense) against the annihilation process[19]. $\blacksquare$

As an immediate consequence, we come to the following

Theorem 5.4 *Let* $\Phi \in \mathcal{W}^*$. *Then* $\rho \in \mathcal{D}^*$ *is a kernel of its coherent state representation:*

$$\Phi = \int_{E_{\mathbf{C}}^*} \rho(z)\phi_z\,\nu(dz)$$

if and only if

$$\rho = \rho_1 + \rho_2, \qquad \rho_1 \in \mathcal{D}_{\mathrm{AH}}^*, \quad \rho_2 \in \mathcal{D}_{\mathrm{AH}}^{\perp},$$

with $\rho_1(z) = S\Phi(\bar{z})$.

Acknowledgments

This work is supported in part by Grant-in-Aid for Scientific Research No. 12440036, Ministry of Education, Japan.

Appendix: Multi-Variable White Noise Functions

In the proof of Theorem 4.3 we used a result on white noise functions of two variables. More generally, taking the n-fold tensor power of the one-variable CKS-space

$$\mathcal{W} \subset L^2(E^*, \mu) \subset \mathcal{W}^*, \qquad E^* = \mathcal{S}'(\mathbf{R}), \tag{50}$$

we obtain a Gelfand triple of n-variable functions:

$$\mathcal{W}^{\otimes n} \subset L^2(\overbrace{E^* \times \cdots \times E^*}^{n\ \text{times}}, \overbrace{\mu \times \cdots \times \mu}^{n\ \text{times}}) \subset (\mathcal{W}^{\otimes n})^*. \tag{51}$$

In this Appendix we mention characterization theorems for (51). In fact, owing to nuclearity of $\mathcal{W}$, most results concerning the one-variable CKS-space (50) admit straightforward generalization to the (finitely many) multi-variable case (51). However, for its importance in applications and lack of written literatures it seems worthwhile to give a brief account.

Throughout this section we fix an integer $n \geq 1$ and consider a finite set $T = \{1, 2, \ldots, n\}$ equipped with the counting measure. By the standard construction[18] with the help of the operator $A \otimes 1$ on $L^2(\mathbf{R} \times T) \cong L^2(\mathbf{R}) \otimes \mathbf{R}^{\otimes n}$, where $A = 1 + t^2 - d^2/dt^2$ as in Section 1.2, we obtain a real Gelfand triple:

$$\mathcal{S}(\mathbf{R} \times T) \subset L^2(\mathbf{R} \times T) \subset \mathcal{S}'(\mathbf{R} \times T). \tag{52}$$

Using nuclearity of $\mathcal{S}(\mathbf{R})$, one sees easily that

$$\mathcal{S}(\mathbf{R} \times T) \cong E \otimes \mathbf{R}^n \cong \overbrace{E \oplus \cdots \oplus E}^{n\text{ times}}, \qquad E = \mathcal{S}(\mathbf{R}), \tag{53}$$

and hence

$$\mathcal{S}'(\mathbf{R} \times T) \cong E^* \otimes \mathbf{R}^n \cong \overbrace{E^* \oplus \cdots \oplus E^*}^{n\text{ times}}, \qquad E^* = \mathcal{S}'(\mathbf{R}). \tag{54}$$

We denote elements of $\mathcal{S}(\mathbf{R} \times T)$ and $\mathcal{S}'(\mathbf{R} \times T)$ by $\boldsymbol{\xi} = (\xi_1, \ldots, \xi_n)$ and $\boldsymbol{x} = (x_1, \ldots, x_n)$, respectively. Then

$$\langle \boldsymbol{x}, \boldsymbol{\xi} \rangle = \sum_{k=1}^{n} \langle x_k, \xi_k \rangle .$$

Here it is important to note that the direct sum $E \oplus \cdots \oplus E$ is obtained from a single space $\mathbf{R} \times T$ and a single operator $A \otimes 1 \cong A \oplus \cdots \oplus A$ acting on $L^2(\mathbf{R} \times T) \cong L^2(\mathbf{R}) \otimes \mathbf{R}^n \cong L^2(\mathbf{R}) \oplus \cdots \oplus L^2(\mathbf{R})$ (n times).

Let $\widetilde{\mu}$ be the standard Gaussian measure associated with the Gelfand triple (52), i.e., determined by

$$e^{-|\boldsymbol{\xi}|_0^2/2} = \int_{\mathcal{S}'(\mathbf{R} \times T)} e^{i\langle \boldsymbol{x}, \boldsymbol{\xi} \rangle} \widetilde{\mu}(d\boldsymbol{x}), \qquad \boldsymbol{\xi} \in \mathcal{S}(\mathbf{R} \times T). \tag{55}$$

Recall that the one-variable CKS-space (50) is constructed from $L^2(E^*, \mu)$ with the help of p-norms and a given weight sequence α. In the multivariable case, the p-norm is defined by the second quantized operator $\Gamma(A \otimes 1)$ and in the exactly same manner we come to the Gelfand triple:

$$\widetilde{\mathcal{W}} \subset L^2(\mathcal{S}'(\mathbf{R} \times T), \widetilde{\mu}) \cong \Gamma(L^2(\mathbf{R} \times T)) \subset \widetilde{\mathcal{W}}^*. \tag{56}$$

On the other hand, by the uniqueness of the characteristic function, we see from (55) that $\tilde{\mu} = \mu \times \cdots \times \mu$ (n times) according to the identification (54). Hence

$$L^2(\mathcal{S}'(\mathbf{R} \times T), \tilde{\mu}) \cong L^2(E^*, \mu)^{\otimes n}.$$

Then by nuclearity of $\mathcal{W}$ one sees from (51) that

$$\widetilde{\mathcal{W}} \cong \mathcal{W}^{\otimes n}.$$

In other words, the Gelfand triple (51) coincides exactly with the CKS-space constructed from $L^2(\mathbf{R} \times T)$.

Since structure of the underlying space $\mathbf{R}$ has no effect on the proofs of characterization theorems for the S-transform and for the operator symbol, the same argument can be applied in the case where the underlying space is $\mathbf{R} \times T$. We only note that

$$\| (A \otimes 1)^{-q} \|_{\mathrm{OP}} = \| A^{-q} \|_{\mathrm{OP}}, \qquad \| (A \otimes 1)^{-q} \|_{\mathrm{HS}}^2 = n \| A^{-q} \|_{\mathrm{HS}}^2.$$

Thus, as literal translation from Theorems 3.1 and 4.1 (the case of $n = 1$) we obtain the following

Theorem A.1 *A function $F : \mathcal{S}_{\mathbf{C}}(\mathbf{R} \times T) \to \mathbf{C}$ is the S-transform of some $\Phi \in (\mathcal{W}^{\otimes n})^*$ if and only if*

(F1) $z \mapsto F(z\boldsymbol{\xi} + \boldsymbol{\xi}')$ *is entire holomorphic in $z \in \mathbf{C}$ for any $\boldsymbol{\xi}, \boldsymbol{\xi}' \in \mathcal{S}_{\mathbf{C}}(\mathbf{R} \times T)$;*

(F2) *there exist $C \geq 0$ and $p \geq 0$ such that*

$$|F(\boldsymbol{\xi})|^2 \leq C G_\alpha(|\boldsymbol{\xi}|_p^2), \qquad \boldsymbol{\xi} \in \mathcal{S}_{\mathbf{C}}(\mathbf{R} \times T).$$

Moreover, in that case,

$$\| \Phi \|_{-(p+q),-}^2 \leq C \widetilde{G}_\alpha(n \| A^{-q} \|_{\mathrm{HS}}^2), \tag{57}$$

where $q > 1/2$ is taken in such a way that $\widetilde{G}_\alpha(n \| A^{-q} \|_{\mathrm{HS}}^2) < \infty$.

Theorem A.2 *A function $\Theta : \mathcal{S}_{\mathbf{C}}(\mathbf{R} \times T) \times \mathcal{S}_{\mathbf{C}}(\mathbf{R} \times T) \to \mathbf{C}$ is the symbol of an operator $\Xi \in \mathcal{L}(\mathcal{W}^{\otimes n}, (\mathcal{W}^{\otimes n})^*)$ if and only if*

(O1) *for any $\boldsymbol{\xi}, \boldsymbol{\xi}', \boldsymbol{\eta}, \boldsymbol{\eta}' \in \mathcal{S}_{\mathbf{C}}(\mathbf{R} \times T)$ the function $(z, w) \mapsto \Theta(z\boldsymbol{\xi} + \boldsymbol{\xi}', w\boldsymbol{\eta} + \boldsymbol{\eta}')$ is entire holomorphic on $\mathbf{C} \times \mathbf{C}$;*

(O2) *there exist constant numbers $C \geq 0$ and $p \geq 0$ such that*

$$|\Theta(\boldsymbol{\xi}, \boldsymbol{\eta})|^2 \leq C G_\alpha(|\boldsymbol{\xi}|_p^2) G_\alpha(|\boldsymbol{\eta}|_p^2), \qquad \boldsymbol{\xi}, \boldsymbol{\eta} \in \mathcal{S}_{\mathbf{C}}(\mathbf{R} \times T).$$

In that case

$$\| \Xi\phi \|^2_{-(p+q),-} \leq C\widetilde{G}^2_\alpha(n\| A^{-q} \|^2_{\mathrm{HS}}) \| \phi \|^2_{p+q,+}, \qquad \phi \in \mathcal{W}^{\otimes n}, \tag{58}$$

where $q > 1/2$ is taken as $\widetilde{G}_\alpha(n\| A^{-q} \|^2_{\mathrm{HS}}) < \infty$.

Characterizations for $\mathcal{W}^{\otimes n}$ and for $\mathcal{L}(\mathcal{W}^{\otimes n}, \mathcal{W}^{\otimes n})$ are also literal translation from known results for the case of $n = 1$. Instead of the translations we shall give more practical statements.

We change notation. An exponential vector for (56) is given by

$$\phi_\xi(x) = e^{\langle x, \xi \rangle - \langle \xi, \xi \rangle/2}, \qquad \xi \in \mathcal{S}_{\mathbf{C}}(\mathbf{R} \times T), \quad x \in \mathcal{S}'_{\mathbf{C}}(\mathbf{R} \times T). \tag{59}$$

With the identification $\xi = (\xi_1, \ldots, \xi_n) \in \mathcal{S}(\mathbf{R} \times T)$ and $x = (x_1, \ldots, x_n) \in \mathcal{S}'(\mathbf{R} \times T)$, (59) becomes

$$\phi_\xi(x) = \prod_{k=1}^n e^{\langle x_k, \xi_k \rangle - \langle \xi_k, \xi_k \rangle/2} = \prod_{k=1}^n \phi_{\xi_k}(x_k).$$

Hence, the S-transform of $\Phi \in (\mathcal{W}^{\otimes n})^*$ is identified with a function on $E_{\mathbf{C}}^n$ in such a way that

$$S\Phi(\xi_1, \ldots, \xi_n) = \langle\!\langle \Phi, \phi_{\xi_1} \otimes \cdots \otimes \phi_{\xi_n} \rangle\!\rangle, \qquad \xi_1, \ldots, \xi_n \in E_{\mathbf{C}}.$$

Similarly, the symbol of $\Xi \in \mathcal{L}(\mathcal{W}^{\otimes n}, (\mathcal{W}^{\otimes n})^*)$ is defined by

$$\widehat{\Xi}(\xi_1, \ldots, \xi_n; \eta_1, \ldots, \eta_n) = \langle\!\langle \Xi(\phi_{\xi_1} \otimes \cdots \otimes \phi_{\xi_n}), \phi_{\eta_1} \otimes \cdots \otimes \phi_{\eta_n} \rangle\!\rangle.$$

After this change of notation Theorems A.1 and A.2 are paraphrased. We only need to recall Hartogs' theorem of holomorphy and multiplicative properties of G_α described in Proposition 1.2. It is convenient to say that a $\mathbf{C}$-valued function F defined on $\mathcal{X}^n$, where $\mathcal{X}$ is a complex topological vector space, is *G-entire (Gâteaux-entire)* if $z \mapsto F(\xi_1, \ldots, \xi_k + z\xi', \ldots \xi_n)$ is entire holomorphic in $z \in \mathbf{C}$ for any $\xi_1, \ldots, \xi_n, \xi' \in \mathcal{X}$ and $1 \leq k \leq n$.

Theorem A.3 *A function $F : E_{\mathbf{C}}^n \to \mathbf{C}$ is the S-transform of some $\Phi \in (\mathcal{W}^{\otimes n})^*$ if and only if*

(F1) *F is G-entire;*

(F2) *there exist $C \geq 0$ and $p \geq 0$ such that*

$$|F(\xi_1, \ldots, \xi_n)|^2 \leq C \prod_{k=1}^n G_\alpha(| \xi_k |^2_p), \qquad \xi_1, \ldots, \xi_n \in E_{\mathbf{C}}.$$

Theorem A.4 *A function $\Theta : E_{\mathbf{C}}^n \times E_{\mathbf{C}}^n \to \mathbf{C}$ is the symbol of an operator $\Xi \in \mathcal{L}(\mathcal{W}^{\otimes n}, (\mathcal{W}^{\otimes n})^*)$ if and only if*

(O1) Θ *is G-entire;*

(O2) *there exist constant numbers $C \geq 0$ and $p \geq 0$ such that*

$$|\Theta(\xi_1,\ldots,\xi_n;\eta_1,\ldots,\eta_n)|^2 \leq C \prod_{k=1}^{n} G_\alpha(|\xi_k|_p^2)G_\alpha(|\eta_k|_p^2),$$

for $\xi_1,\ldots,\xi_n,\eta_1,\ldots,\eta_n \in E_{\mathbf{C}}$.

Theorem A.5 *A function $F : E_{\mathbf{C}}^n \to \mathbf{C}$ is the S-transform of some $\Phi \in \mathcal{W}^{\otimes n}$ if and only if*

(F1) *F is G-entire;*

(F2) *for any $p \geq 0$ there exists $C \geq 0$ such that*

$$|F(\xi_1,\ldots,\xi_n)|^2 \leq C \prod_{k=1}^{n} G_{1/\alpha}(|\xi_k|_{-p}^2), \qquad \xi_1,\ldots,\xi_n \in E_{\mathbf{C}}.$$

Theorem A.6 *A function $\Theta : E_{\mathbf{C}}^n \times E_{\mathbf{C}}^n \to \mathbf{C}$ is the symbol of an operator $\Xi \in \mathcal{L}(\mathcal{W}^{\otimes n}, \mathcal{W}^{\otimes n})$ if and only if*

(O1) Θ *is G-entire;*

(O2) *for any $p \geq 0$ there exist $C \geq 0$ and $q \geq 0$ such that*

$$|\Theta(\xi_1,\ldots,\xi_n;\eta_1,\ldots,\eta_n)|^2 \leq C \prod_{k=1}^{n} G_\alpha(|\xi_k|_{p+q}^2)G_{1/\alpha}(|\eta_k|_{-p}^2),$$

for $\xi_1,\ldots,\xi_n,\eta_1,\ldots,\eta_n \in E_{\mathbf{C}}$.

References

1. S. T. Ali, J.-P. Antoine and J.-P. Gazeay: "Coherent States, Wavelets and Their Generalizations," Springer–Verlag, 2000.
2. N. Asai, I. Kubo and H.-H. Kuo: *General characterization theorems and intrinsic topologies in white noise analysis*, to appear in Hiroshima Math. J. **31** (2001).
3. Y. M. Berezansky and Y. G. Kondratiev: "Spectral Methods in Infinite-Dimensional Analysis," Kluwer Academic Publisher, 1995.
4. D. M. Chung, U. C. Ji and N. Obata: *Higher powers of quantum white noises in terms of integral kernel operators*, Infinite Dimen. Anal. Quantum Probab. Rel. Top. **1** (1998), 533–559.

5. D. M. Chung, U. C. Ji and N. Obata: *Quantum stochastic analysis via white noise operators in weighted Fock space*, to appear in Rev. Math. Phys.

6. W. G. Cochran, H.-H. Kuo and A. Sengupta: *A new class of white noise generalized functions*, Infinite Dimen. Anal. Quantum Probab. Rel. Top. **1** (1998), 43–67.

7. R. Gannoun, R. Hachaichi, H. Ouerdiane and A. Rezgui: *Un théoreme de dualité entre espaces de fonctions holomorphes à croissance exponentielle*, J. Funct. Anal. **171** (2000), 1–14.

8. L. Gross and P. Malliavin: *Hall's transform and the Segal–Bargmann map*, in "Itô's Stochastic Calculus and Probability Theory (N. Ikeda, S. Watanabe, M. Fukushima and H. Kunita (Eds.)," pp. 73–116, Springer–Verlag, 1996.

9. T. Hida: "Brownian Motion," Springer–Verlag, 1980.

10. U. C. Ji and N. Obata: *A role of Bargmann–Segal spaces in characterization and expansion of operators on Fock space*, preprint, 2000.

11. U. C. Ji, N. Obata and H. Ouerdiane: *Analytic characterization of generalized Fock space operators as two-variable entire functions with growth condition*, to appear in Infinite Dimen. Anal. Quantum Probab. Rel. Top., 2001.

12. J. R. Klauder: *The action option and a Feynman quantization of spinor fields in terms of ordinary C-numbers*, Ann. Phys. **11** (1960), 123–168.

13. J. R. Klauder and B.-S. Skagerstam: "Coherent States," World Scientific, 1985.

14. I. Kubo and S. Takenaka: *Calculus on Gaussian white noise I–IV*, Proc. Japan Acad. **56A** (1980), 376–380; 411–416; **57A** (1981), 433–437; **58A** (1982), 186–189.

15. I. Kubo and Y. Yokoi: *Generalized functions and functionals in fluctuation analysis*, in "Mathematical Approach to Fluctuations II (T. Hida, Ed.)," pp. 203–230, World Scientific, 1995.

16. H.-H. Kuo: "White Noise Distribution Theory," CRC Press, 1996.

17. Y.-J. Lee: *Analytic version of test functionals, Fourier transform and a characterization of measures in white noise calculus*, J. Funct. Anal. **100** (1991), 359–380.

18. N. Obata: "White Noise Calculus and Fock Space," Lect. Notes in Math. Vol. 1577, Springer–Verlag, 1994.

19. N. Obata: *Integral kernel operators on Fock space – Generalizations and applications to quantum dynamics*, Acta Appl. Math. **47** (1997), 49–77.

20. N. Obata: *A note on Hida's whiskers and complex white noise*, in "Analysis on Infinite Dimensional Lie Groups and Algebras (H. Heyer and J.

Marion, Eds.)," pp. 321–336, World Scientific, 1998.

21. N. Obata: *Coherent state representations in white noise calculus*, Can. Math. Soc. Conference Proceedings **29** (2000), 517–531.

22. N. Obata: *Coherent state representation and unitarity condition in white noise calculus*, J. Korean Math. Soc. **38** (2001), 297–309.

23. N. Obata: *Unitarity criterion in white noise calculus and nonexistence of unitary evolutions driven by higher powers of quantum white noises*, to appear in Proc. Mexico Guanajuato Conference Proceedings, 2001.

24. H. Ouerdiane: *Fonctionnelles analytiques avec condition de croissance, et application à l'analyse gaussienne*, Japan. J. Math. **20** (1994), 187–198.

25. J. Potthoff and L. Streit: *A characterization of Hida distributions*, J. Funct. Anal. **101** (1991), 212–229.

26. Y. Yokoi: *Simple setting for white noise calculus using Bargmann space and Gauss transform*, Hiroshima Math. J. **25** (1995), 97–121.

Quantum Information IV (pp. 177–185)
Eds. T. Hida and K. Saitô
© 2002 World Scientific Publishing Co.

TOPICS ON COMPLEX GAUSSIAN RANDOM FIELDS

SI SI

Faculty of Information Science and Technology
Aichi Prefectural University
Aichi-ken 480-1198, Japan

WIN WIN HTAY

Department of Computational Mathematics
University of Computer Studies
Yangon, Myanmar

Mathematics Subject Classification (2000): 60H40

0 Introduction

We are interested in the complex random field which can be expressed in terms of complex white noise as

$$Z(C) = \int_{(C)^n} F(C;u)z(u_1)\cdots z(u_n)du^n,$$

where C runs through the class

$$\mathbf{C}^{d-1} = \{C; C \text{ diffeomorphic to } S^{d-1}, \text{ convex }\}.$$

The $z(u), u \in R^d$, denotes the complex white noise: $x(u) + iy(u)$, where $x(u)$ and $y(u)$ are mutually independent real white noises.

It is observed that the complex random field $Z(C)$ is harmonic in the sense of the Gross Laplacian, and further the innovation of $Z(C)$ is established by taking its variation.

1 Background

We are going to discuss complex Gaussian random fields given by

$$Z(C) = \int_{(C)^n} F(C;u)z(u_1)\cdots z(u_n)du^n, \tag{1.1}$$

which is parameterized by C in $\mathbf{C^{d-1}}$:

$$\mathbf{C^{d-1}} = \{C;\ \text{diffeomorphic to } S^{d-1},\ \text{convex}\},$$

where (C) denotes the domain enclosed by C. The $z(u), u \in R^d$ denotes the d-dimensional parameter complex white noise which is expressible as $z(u) = x(u) + iy(u)$, x , y being mutually independent real white noises.

As the background, we first recall the complex Gaussian systems and the complex white noise.

1.1 Complex Gaussian System

Let $(\Omega, \mathcal{B}, P)$ be a probability space and let $Z(\omega)$ be a complex-valued random variable on $(\Omega, \mathcal{B}, P)$, such that $E(X(\omega)) = m$, $m \in C($ complex numbers $)$, $Re(Z(\omega)) = X(\omega)$ and $Im(Z(\omega)) = Y(\omega)$.

Definition If $X(\omega)$ and $Y(\omega)$ are independent and they have the same Gaussian distribution with mean zero, then

$$Z(\omega) = m + X(\omega) + iY(\omega)$$

is called *complex Gaussian*.

Definition Let $\mathbf{Z} = \{Z_\lambda(\omega); \lambda \in \Lambda\}$ be a system of complex-valued random variables. If any linear combination $\sum_j c_j Z_{\lambda_j}$, $c_j \in C$, is complex Gaussian, then $\mathbf{Z}$ is called a complex Gaussian system.

Theorem *For any vector $\mathbf{m} = (m_\lambda; \lambda \in \Lambda)$ and positive definite matrix $\mathbf{V} = (V_{\lambda,\mu}; \lambda, \mu \in \Lambda)$ on Λ, there exists a complex Gaussian system $\mathbf{Z} = \{Z_\lambda : \lambda \in \Lambda\}$ with a mean vector and a covariance matrix which coincides with the given vector $\mathbf{m}$ and the matrix $\mathbf{V}$, respectively.*

Let the random variable $Z(\omega)$ be complex Gaussian, $z \in C$ be a complex variable, and t and s be arbitrary complex numbers. We have

1. $E[\exp i(tZ + s\bar{Z})] = \exp[i(tm + s\overline{m}) - ts\sigma^2]$, where $m = E(Z)$ and $\sigma^2 = E[|Z - m|^2]$.

2. If $E[Z] = 0$, then we have

$$E[\exp(tZ)] = 1,\ E[Z^n] = 0,\ n > 0,\ E[Z^n \bar{Z}^m] = \delta_{n,m} n! E[\{|Z|^2\}^n].$$

3. Let $H_{p,q}(z,\bar{z}) = (-1)^{p+q}e^{|z|^2}\frac{\partial^{p+q}}{\partial\bar{z}^p\partial z^q}e^{-|z|^2}$, $\quad p,q \geq 0$, be the *complex Hermite polynomial of degree* (p,q). Then, the system of functions of $Z(\omega)$

$$\{(p!q!)^{-\frac{1}{2}}H_{p,q}(Z(\omega),\bar{Z}(\omega)); p,q \geq 0\},$$

is an orthonomal system in $L^2(\Omega, P)$. (See e.g. [2].)

1.2 Complex white noise

Let $E_c = E + iE$ and $E_c^* = E^* + iE^*$ with a nuclear space $E(\subset L^2(R^d))$. Denote $\zeta \in E_c$ by $\zeta = \xi + i\eta$, $\xi, \eta \in E$. Then, $z \in E_c^*$ is of the form $z = x + iy$, $x, y \in E^*$.

Set $\nu = \mu_1 \times \mu_2$, where μ_1 and μ_2 are white noise measures with variance $\frac{1}{2}$ on $E^*(\cong iE^*)$ such that

$$\int_{E^*} \exp[i\langle x, \xi\rangle]d\mu_j(x) = \exp\left[-\frac{1}{4}\|\xi\|^2\right], \ j = 1, 2.$$

Let $\mathcal{B}$ be the smallest sigma field generated by the cylinder subsets of E_c^*. Then, $(E_c^*, \mathcal{B}, \nu)$ is called a *complex white noise*. The class

$$\mathcal{H}^0 \equiv \{\langle z, \zeta\rangle; \zeta \in E_C\}$$

is a complex Gaussian system on $(E_c^*, \mathcal{B}, \nu)$, and so is $\overline{\mathcal{H}^0} \equiv \{\overline{\langle z, \zeta\rangle}; \zeta \in E_c\}$.
We have

$$E[\langle z, \zeta\rangle] = \int_{E_C^*} \langle z, \zeta\rangle d\nu(z) = 0, \ E[\langle z, \zeta\rangle^2] = \|\zeta\|^2,$$

where $\|\cdot\|$ denotes the $L^2(R)$-norm. The covariance matrix of $\mathcal{H}^0$ is given by

$$E[\langle z, \zeta_1\rangle\overline{\langle z, \zeta_2\rangle}] = \langle\zeta_1, \bar{\zeta_2}\rangle.$$

1.3. Complex multiple Wiener integral of degree (m,n)

Denote the subspace of (L_c^2) spanned by the Fourier-Hermite polynomials of degree (m, n) based upon a complete orthonormal system $\{w_n\}$ by $\mathcal{H}_{(m,n)}$. Some known results are now in order.

1. The family of all functionals of the form

$$\prod_k (p!q!)^{-\frac{1}{2}}H_{p,q}(\langle z, w_{j_k}\rangle, \overline{\langle z, w_{j_k}\rangle});$$

where the product is finite and the j_k's are distinct, is a complete orthonomal system in (L_c^2).

2. An element of $\mathcal{H}_{(m,n)}$ is now called a complex multiple Wiener integral of degree (m, n). Set

$$\mathcal{H}_n = \sum_{k=0}^{n} \bigoplus \mathcal{H}_{(n-k,k)},$$

which is the collection of members of degree n. The space (L_c^2) has therefore the direct sum decomposition

$$(L_c^2) = \sum_{n=0}^{\infty} \bigoplus \left(\sum_{k=0}^{n} \bigoplus \mathcal{H}_{(n-k,k)} \right) = \sum_{n=0}^{\infty} \bigoplus \mathcal{H}_n$$

3. A member $\varphi(z)$ in $\mathcal{H}_{(n,0)}$ has an integral representation of the form

$$\varphi(z) = \int_{R^n} F(u)z(u_1) \cdots z(u_n)du^n,$$

where F is a symmetric $L^2(R^n)$-function. (For simplicity d is taken to be 1.) A generalization of such a representation can be obtained for any $\mathcal{H}_{(m,n)}$-functional.

2 Complex Gaussian random field

Define a complex Gaussian random field $Z(C)$ by

$$Z(C) = \int_{(C)} F(C;u)z(u_1) \cdots z(u_n)du^n, \tag{2.1}$$

where

$$\mathbf{C}^{d-1} = \{C; C \in C^2, \text{diffeomorphic to } S^{d-1}\}, \tag{2.2}$$

and $z = x + iy$ is a complex white noise. Let

(i) Let L_C be the closed sbspace of $L^2(E_c^*, \nu)$ generated by the algebra involving the $\langle z, \zeta \rangle$, $supp(\zeta) \subset (C)$.

(ii) the sigma field $\mathcal{B}_{(C)}$ is the smallest sigma field of subsets of E_c^* with respect to which all members in L_C are measurable.

Obviously $Z(C)$ lives in the space L_C and

$$L_C \subset L^2(\mathcal{B}_{(C)}, \nu) \subset (L_c^2).$$

Proposition. Set $L = \bigvee_C L_C$ (the lattice sum). Then, we have a direct sum decomposition of the space L:

$$L = \bigoplus_n \mathcal{H}_{(n,0)}.$$

As in [3], we can introduce the differential operators where the white noises $x(u)$ and $y(u)$ are taken to be variables. Namely,

$$\frac{\partial}{\partial x(u)}, \ \frac{\partial}{\partial y(u)}$$

as well as the higher order partial derivatives. Indeed, they can be defined rigorously by using our standard technique of white noise analysis, although one might think the variables have a formalsignificance.

Moreover, we can introduce the Gross Laplacian Δ_G with or without restriction of the domain of the parameter u. For instance,

$$\Delta_G(C) = \int_{(C)} [\frac{\partial^2}{\partial x(u)^2} + \frac{\partial^2}{\partial y(u)^2}] du^d.$$

By expressing z as $x + iy$, we can easily see that

$$\left(\frac{\partial}{\partial x(u)}\right)^2 Z(C) + \left(\frac{\partial}{\partial y(u)}\right)^2 Z(C) = 0$$

holds for $Z(C)$ given by (2.1). Thus, we have

$$(\Delta_G)(C)(Z(C)) = 0.$$

With a particular choice of $(C) = R^d$ we can define $\Delta_G(R^d)$ which is simply denoted by Δ_G.

Theorem 1. *For any C the random field $Z(C) = Z(C, z)$ is harmonic in the sense of the Gross Laplacian Δ_G.*

Remark Before closing this section we shall make a short, general remark on the representation of a stochastic process or a random field which is expressed as a stochastic integral with respect to a random measure. The notion of the canonical kernel was first introduced when the canonical representation of a Gaussian process was introduceed. This particular property of a representation means that, intuitively speaking, the random measure and the represented process have exactly the same information until any fixed time. Hence, there is no need to restrict the process to be Gaussian to define the canonical property of the kernel. With this understanding, we have the freedom to use the established criterion for the kernel to be canonical, in the case where the given process or the random field is expressed as a stochastic integral.

With this remark, we assume that the kernel F in the integral of the form (2.1) is always taken to be canonical, in what follows.

3 Innovation

It is known that the infinitesimal equation (the stochastic variational equation) for a real Gaussian random field $X(C)$ can be defined as a generalization of the Lévy's infinitesimal equation for the variation $\delta X(t)$ of a stochastic process $X(t)$. For the case of a random field $X(C)$, the equation for the variation $\delta X(C)$ may be expressed in the form

$$\delta X(C) = \Phi(X(C')\ C' < C, Y(s), s \in C, C, \delta C), \tag{3.1}$$

where $C' < C$ means that C' is inside of C, that is, the domain (C') enclosed by C' is a subset of (C), and where Φ is some nonrandom function. The system $Y = \{Y(s), s \in C\}$ is the *innovation* in the sense that $Y(s)$ is independent of the past $\{X(C'), C' < C\}$ and that it contains the full information that $X(C)$ gains by taking the infinitesimal change of C.

Remind the definition of $\mathbf{C^{d-1}}$. With this choice of C in that class, we can guarantee the existence of a complex white noise $z(s)$ with parameter $s \in C$ and such white noises with various C forms a consistent family when C moves around in R^d being restricted to be in $\mathbf{C^{d-1}}$..

The infinitesimal equation for the *complex* Gaussian random field $X(C)$ can be introduced in a similar manner:

$$\delta Z(C) = \Phi(Z(C'); C' < C, z(s), \overline{z(s)}, s \in C, C, \delta C),$$

where $C' < C$ means, as before, that C' is inside of C, and where Φ is a nonrandom function. The system $\{z(s), \overline{z(s)}, s \in C; C\}$ is the innovation in the sense that the family of the pairs $\{z(s), z(s), s \in C, \}$ is independent of the past $\{Z(C'), C' < C\}$ and that it describes the new information gained by $Z(C)$ in the infinitesimal neighbourhood of C.

Remark. For the innovation it suffices for us to obtain either one of $z(s)$ and $\overline{z(s)}$ from $\delta Z(C)$, since the other is complex conjugate.

Theorem 2. *The innovation for the random field $Z(C)$ given by (3.2) is*

$$z(s) = \frac{1}{\varphi(s)}\{\frac{\delta Z(C) - \hat{E}(\delta Z(C)|Z(C'), C' < C)}{\delta n}(s)\}, \quad s \in C, \qquad (3.2)$$

where $\varphi(s)$ is a random function of s determined by the weak conditional expectation of $X(C)$.

Proof. Take the variation $\delta Z(C)$ of the random field $Z(C)$ defined by (3.2). Then, it can be written in the form

$$\delta Z(C) = n \int_C \int_{(C)^{n-1}} F(C, v_1; s) z^{(n-1)\otimes}(v_1) z(s) dv_1^{n-1} \delta n(s) ds$$

$$+ \int_C \int_{(C)^n} F'_n(C, u; s) z^{n\otimes}(u) \delta n(s) du^n ds, \qquad (3.3)$$

where $v_1 = (u_1, u_2, \cdots, u_{n-1})$ and F'_n denotes the functional derivative of $F(C; u)$ evaluated at C. Now take the weak conditional expectation $\hat{E}$ which

is defined by the orthogonal projection:

$$\hat{E}(\delta Z(C)|Z(C'), C' < C) = \int_C \int_{(C)^n} F'_n(C, u; s) z^{n\otimes}(u) \delta n(s) du^n ds.$$

Then, we have

$$\delta Z(C) - \hat{E}(\delta Z|Z(C'), C' < C)$$

$$= n \int_C \int_{(C)^{n-1}} F(C, v_1; s) z^{(n-1)\otimes}(v_1) z(s) dv_1^{n-1} \delta n(s) ds. \qquad (3.4)$$

Let δn vary all over in the class of positive C^∞–functions in such a way that δC is taken outward and that the integrand over C is determined as a function of the variable s. Then, the right hand side of (3.5) will determine the following

$$z(s) \int_{(C)^{n-1}} F(C, v_1; s) z^{(n-1)\otimes}(v_1) dv_1^{n-1}. \tag{3.5}$$

Let it be denoted as

$$z(s)\varphi(s) \tag{3.6}$$

and use the same technique as in one dimensional parameter cace (cf. [4]). Thus, we know the value

$$\varphi(s)^2. \tag{3.7}$$

We may ignore its sign to determine $\varphi(s)$. Divide (3.7) by $\varphi(s)$ to obtain the generalized innovation $z(s)$ for a moment. Since the representation is cannonical, it can be regarded as the same as the original complex white noise $z(s)$. This means that it is the real innovation (not in a generalized sense). The Theorem is therefore proved.

It should be noted that the equation (3.2) gives the original complex white noise z in (2.1), because the kernel F is canonical. However, if the kernel F is not cannonical, we can see an example such that the following inequality holds.

$$\delta Z(C) - \hat{E}(\delta Z | Z(C'), C' < C) \neq$$
$$n \int_C \int_{(C)^{n-1}} F(C, v_1; s) z^{(n-1)\otimes}(v_1) z(s) dv_1^{n-1} \delta n(s) ds. \tag{3.8}$$

Hence, in such a case we are given a generalized innovation which may be different from the original z.

Remark. The innovation $\overline{z(s)}$ can not be obtained in L_C, however it can exist in the space (L_c^2).

Now, as an appendix, we wish to note some elementary, but useful relationships among the real stochastic processes arising from the Fourier transform of white noise.

Take a real white noise $x(t)$. Then, the Fourier transform $\hat{x}(\lambda)$ of $x(t)$ can be expressed in the form

$$\hat{x}(\lambda) = \hat{x}_r(\lambda) + i\hat{x}_i(\lambda) \tag{3.9}$$

where $\hat{x}_r(\lambda)$ and $\hat{x}_i(\lambda)$ are real valued. Then, we have *formal* expressions of the form

$$\hat{x}_r(\lambda) = \frac{1}{\sqrt{2\pi}} \int_{-\infty}^{\infty} \cos \lambda u x(u) du \tag{3.10}$$

$$\hat{x}_i(\lambda) = \frac{1}{\sqrt{2\pi}} \int_{-\infty}^{\infty} \sin \lambda u x(u) du. \tag{3.11}$$

Thus, we see that

$$\overline{\hat{x}(\lambda)} = \hat{x}(-\lambda). \tag{3.12}$$

Fact. $\{\hat{x}_r(\lambda), \lambda \geq 0\}$ and $\{\hat{x}_i(\lambda), \lambda \geq 0\}$ are (real) white noises and $\{\hat{x}(\lambda), \lambda \geq 0\}$ is a complex white noise.

Acknowledgements The authors are grateful to the organizers of the Quantum Information Conference who gave the opportunity to present the results reported in this paper.

References

1. T. Hida, Cannonical representation of Gaussian processes and their applications, Mem. Coll. Sci. Univ. Kyoto, 33 (1960), 258-351.
2. T. Hida, Brownian motion, Springer-Verlag. 1980.
3. T. Hida et al. White noise. An infinite dimensional calculus, Kluwer Academic Pub. 1993.
4. T. Hida and Si Si, Innovations for random fields. Infinite Dimensional Analysis, Quantum Probability and Related Topics Vol 1, No. 4, World Scientific Pub. Co. (1998), 499-510.
5. Si Si, Gaussian processes and Gaussian random fields. Quantum Information II, ed. by T.Hida and K. Saito, World Scientific Pub. Co. 2000, 195-204.
6. Si Si, Random fields and multiple Markov properties. Supplementary Papers for the Second International Conference on Unconventional Models of Computation UMC'2K, CDMTCS 147, Solvay Institute 2000, 64-70.

Quantum Information IV (pp. 187–200)
Eds. T. Hida and K. Saitô

QUANTUM INFORMATION IN SPACE AND TIME AND THEORY OF STOCHASTIC PROCESSES

IGOR V. VOLOVICH

Steklov Mathematical Institute
Gubkin St.8, 117966 Moscow, Russia
E-mail: volovich@mi.ras.ru

Many important results in modern quantum information theory have been obtained for an idealized situation when the spacetime dependence of quantum phenomena is neglected. We emphasize the importance of the investigation of quantum information in space and time. Entangled states in space and time are considered. A modification of Bell's equation which includes the spacetime variables is suggested. A general relation between quantum theory and theory of classical stochastic processes is proposed which expresses the condition of local realism in the form of a noncommutative spectral theorem. Applications of this relation to the security of quantum key distribution in quantum cryptography are considered.

1 Introduction

Recent remarkable experimental and theoretical results have shown that quantum effects can provide qualitatively new forms of communication and computation, sometimes more powerful then the classical ones, for a review see e.g. [1]. Interesting and important results obtained in quantum computing, teleportation and cryptography are based on the investigation of basic properties of quantum mechanics. Especially important are properties of non-factorized entangled states which were named by Schrodinger as the most characteristic feature of quantum mechanics.

Modern quantum information theory is built on ideas of classical information theory of C. Shannon and on the notions of von Neumann quantum mechanical entropy and of entangled states as formulated by J. Bell, see [2] for a recent review. The spacetime dependence is not explicitly indicated in this approach. As a result, many important achievements in modern quantum information theory have been obtained for an idealized situation when the spacetime dependence of quantum phenomena is neglected. We emphasize the importance of the investigation of quantum information effects in space and time. [a]

---

[a] The importance of the investigation of quantum information effects in space and time and especially the role of relativistic invariance in classical and quantum information theory was stressed in the talk by the author at the round table at the First International Conference on Quantum Information which was held at Meijo University, November 4-8, 1997.

188

In this paper entangled states in space and time are considered. A modification of Bell's equation which includes the spacetime variables is suggested. A general relation between quantum theory and classical theory of stochastic processes is proposed which expresses the condition of local realism in the form of a noncommutative spectral theorem. Applications of this relation to the security of quantum key distribution in quantum cryptography are considered.

Entangled states, i.e. the states of two particles with the wave function which is not a product of the wave functions of single particles, have been studied in many theoretical and experimental works starting from works of Einstein, Podolsky and Rosen, Bohm and Bell, see e.g. [3].

Bell's theorem [4] states that there are quantum spin correlation functions that can not be represented as classical correlation functions of separated random variables. It has been interpreted as incompatibility of the requirement of locality with the statistical predictions of quantum mechanics [4]. For a recent discussion of Bell's theorem see, for example [3] - [9] and references therein. It is now widely accepted, as a result of Bell's theorem and related experiments, that "Einstein's local realism" must be rejected.

Let us note however that, evidently, the very formulation of the problem of locality in quantum mechanics is based on ascribing a special role to the position in ordinary three-dimensional space. It is rather surprising therefore that the space dependence of the wave function is neglected in discussions of the problem of locality in relation to Bell's inequalities. Actually it is the space part of the wave function which is relevant to the consideration of the problem of locality.

We know that the wave function of particle includes not only the spin part but also the part depending on spacetime variables. Recently it was pointed out [9] that in fact the spacetime part of the wave function was neglected in the proof of Bell's theorem. However just the spacetime part is crucial for considerations of property of locality of quantum system. Actually the spacetime part leads to an extra factor in quantum correlations and as a result the ordinary proof of Bell's theorem fails in this case. We present a modification of Bell's equation which includes space and time variables.

We present a criterion of locality (or nonlocality) of quantum theory in a realist model of hidden variables. We argue that predictions of quantum mechanics can be consistent with Bell's inequalities for some Gaussian wave functions and hence Einstein's local realism is restored in this case. Moreover we show that due to the expansion of the wave packet the locality criterion is always satisfied for nonrelativistic particles if regions of detectors are far enough from each other. This result has applications to the security of certain quantum cryptographic protocols.

We will consider an important connection between quantum mechanics and theory of classical stochastic processes. Consider for example an equation

$$\cos(\alpha - \beta) = E\xi_\alpha\eta_\beta$$

where ξ_α and η_β are two random processes [10] and E is the expectation. Bell's theorem states that there exists no solution of the equation for bounded stochastic processes such that $|\xi_\alpha| \leq 1$, $|\eta_\beta| \leq 1$.

The function $\cos(\alpha - \beta)$ describes the quantum mechanical correlation of spins of two entangled particles. It was shown in [9] that if one takes into account the space part of the wave function then the quantum correlation in the simplest case will take the form $g\cos(\alpha - \beta)$ instead of just $\cos(\alpha - \beta)$ where the parameter g describes the location of the system in space and time. In this case one gets a modified equation

$$g\cos(\alpha - \beta) = E\xi_\alpha\eta_\beta$$

One can prove (see below) that if g is small enough then there exists a solution of the modified equation.

It is important to study also the more general question: which class of functions $f(s,t)$ admits a representation of the form

$$f(s,t) = Ex_sy_t$$

where x_s and y_t are bounded stochastic processes and also analogous question for the functions of several variables $f(t_1, ..., t_n)$.

Bell's theorem constitutes an important part in quantum cryptography [11]. It is now generally accepted that techniques of quantum cryptography can allow secure communications between distant parties [12] - [19]. The promise of secure cryptographic quantum key distribution schemes is based on the use of quantum entanglement in the spin space and on quantum no-cloning theorem. An important contribution of quantum cryptography is a mechanism for detecting eavesdropping.

However in certain current quantum cryptography protocols the space part of the wave function is neglected. But just the space part of the wave function describes the behaviour of particles in ordinary real three-dimensional space. As a result such schemes can be secure against eavesdropping attacks in the abstract spin space but could be insecure in the real three-dimensional space. We will discuss how one can try to improve the security of quantum cryptography schemes in space by using a special preparation of the space part of the wave function, see [18].

Note that some problems of quantum teleportation in space have been discussed in [20].

2 Bell's Theorem

In the presentation of Bell's theorem we will follow [9] where one can find also more references. Bell's theorem reads:

$$\cos(\alpha - \beta) \neq E\xi_\alpha\eta_\beta \tag{1}$$

where ξ_α and η_β are two random processes such that $|\xi_\alpha| \leq 1$, $|\eta_\beta| \leq 1$ and E is the expectation. In more details:

Theorem 1. There exists no probability space $(\Lambda, \mathcal{F}, d\rho(\lambda))$ and a pair of stochastic processes $\xi_\alpha = \xi_\alpha(\lambda)$, $\eta_\beta = \eta_\beta(\lambda)$, $0 \leq \alpha, \beta \leq 2\pi$ which obey $|\xi_\alpha(\lambda)| \leq 1$, $|\eta_\beta(\lambda)| \leq 1$ such that the following equation is valid

$$\cos(\alpha - \beta) = E\xi_\alpha\eta_\beta \tag{2}$$

for all α and β.

Here the expectation is

$$E\xi_\alpha\eta_\beta = \int_\Lambda \xi_\alpha(\lambda)\eta_\beta(\lambda)d\rho(\lambda)$$

We say that Eq. (2) has no solutions $(\Lambda, \mathcal{F}, d\rho(\lambda), \xi_\alpha, \eta_\beta)$ with the bound $|\xi_\alpha| \leq 1$, $|\eta_\beta| \leq 1$.

We will prove the theorem below. Let us discuss now the physical interpretation of this result.

Consider a pair of spin one-half particles formed in the singlet spin state and moving freely towards two detectors. If one neglects the space part of the wave function then one has the Hilbert space $\mathbb{C}^2 \otimes \mathbb{C}^2$ and the quantum mechanical correlation of two spins in the singlet state $\psi_{spin} \in \mathbb{C}^2 \otimes \mathbb{C}^2$ is

$$D_{spin}(a, b) = \langle \psi_{spin} | \sigma \cdot a \otimes \sigma \cdot b | \psi_{spin} \rangle = -a \cdot b \tag{3}$$

Here $a = (a_1, a_2, a_3)$ and $b = (b_1, b_2, b_3)$ are two unit vectors in three-dimensional space $\mathbb{R}^3$, $\sigma = (\sigma_1, \sigma_2, \sigma_3)$ are the Pauli matrices,

$$\sigma_1 = \begin{pmatrix} 0 & 1 \\ 1 & 0 \end{pmatrix}, \quad \sigma_2 = \begin{pmatrix} 0 & -i \\ i & 0 \end{pmatrix}, \quad \sigma_3 = \begin{pmatrix} 1 & 0 \\ 0 & -1 \end{pmatrix},$$

$$\sigma \cdot a = \sum_{i=1}^{3} \sigma_i a_i$$

and

$$\psi_{spin} = \frac{1}{\sqrt{2}} \left(\begin{pmatrix} 0 \\ 1 \end{pmatrix} \otimes \begin{pmatrix} 1 \\ 0 \end{pmatrix} - \begin{pmatrix} 1 \\ 0 \end{pmatrix} \otimes \begin{pmatrix} 0 \\ 1 \end{pmatrix} \right)$$

If the vectors a and b belong to the same plane then one can write $-a \cdot b = \cos(\alpha - \beta)$ and hence Bell's theorem states that the function $D_{spin}(a, b)$ Eq. (3) can not be represented in the form

$$P(a,b) = \int \xi(a, \lambda)\eta(b, \lambda)d\rho(\lambda) \tag{4}$$

i.e.

$$D_{spin}(a, b) \neq P(a, b) \tag{5}$$

Here $\xi(a, \lambda)$ and $\eta(b, \lambda)$ are random fields on the sphere, $|\xi(a, \lambda)| \leq 1$, $|\eta(b, \lambda)| \leq 1$ and $d\rho(\lambda)$ is a positive probability measure, $\int d\rho(\lambda) = 1$. The parameters λ are interpreted as hidden variables in a realist theory. It is clear that Eq. (5) can be reduced to Eq. (1).

To prove Theorem 1 we will use the following

Theorem 2. Let f_1, f_2, g_1 and g_2 be random variables on the probability space $(\Lambda, \mathcal{F}, d\rho(\lambda))$ such that

$$|f_i(\lambda)g_j(\lambda)| \leq 1, \quad i, j = 1, 2.$$

Denote

$$P_{ij} = Ef_ig_j, \quad i, j = 1, 2.$$

Then

$$|P_{11} - P_{12}| + |P_{21} + P_{22}| \leq 2.$$

Proof of Theorem 2. One has

$$P_{11} - P_{12} = Ef_1g_1 - Ef_1g_2 = E(f_1g_1(1 \pm f_2g_2)) - E(f_1g_2(1 \pm f_2g_1))$$

Hence

$$|P_{11} - P_{12}| \leq E(1 \pm f_2g_2) + E(1 \pm f_2g_1) = 2 \pm (P_{22} + P_{21})$$

Now let us note that if x and y are two real numbers then

$$|x| \leq 2 \pm y \quad \rightarrow \quad |x| + |y| \leq 2.$$

Therefore taking $x = P_{11} - P_{12}$ and $y = P_{22} + P_{21}$ one gets the bound

$$|P_{11} - P_{12}| + |P_{21} + P_{22}| \leq 2.$$

The theorem is proved.

The last inequality is called the Clauser-Horn-Shimony-Holt (CHSH) inequality. By using notations of Eq. (4) one has

$$|P(a,b) - P(a,b')| + |P(a',b) + P(a',b')| \leq 2 \tag{6}$$

192

for any four unit vectors a, b, a', b'.

Proof of Theorem 1. Let us denote

$$f_i(\lambda) = \xi_{\alpha_i}(\lambda), \quad g_j(\lambda) = \eta_{\beta_j}(\lambda), \quad i, j = 1, 2$$

for some α_i, β_j. If one would have

$$\cos(\alpha_i - \beta_j) = E f_i g_j$$

then due to Theorem 2 one should have

$$|\cos(\alpha_1 - \beta_1) - \cos(\alpha_1 - \beta_2)| + |\cos(\alpha_2 - \beta_1) + \cos(\alpha_2 - \beta_2)| \leq 2.$$

However for $\alpha_1 = \pi/2$, $\alpha_2 = 0$, $\beta_1 = \pi/4, \beta_2 = -\pi/4$ we obtain

$$|\cos(\alpha_1 - \beta_1) - \cos(\alpha_1 - \beta_2)| + |\cos(\alpha_2 - \beta_1) + \cos(\alpha_2 - \beta_2)| = 2\sqrt{2}$$

which is greater than 2. This contradiction proves Theorem 1.

It will be shown below that if one takes into account the space part of the wave function then the quantum correlation in the simplest case will take the form $g\cos(\alpha - \beta)$ instead of just $\cos(\alpha - \beta)$ where the parameter g describes the location of the system in space and time. In this case one can get a representation

$$g\cos(\alpha - \beta) = E\xi_\alpha\eta_\beta \tag{7}$$

if g is small enough. The factor g gives a contribution to visibility or efficiency of detectors that are used in the phenomenological description of detectors.

3   Localized Detectors

In the previous section the space part of the wave function of the particles was neglected. However exactly the space part is relevant to the discussion of locality. The Hilbert space assigned to one particle with spin $1/2$ is $\mathbb{C}^2 \otimes L^2(\mathbb{R}^3)$ and the Hilbert space of two particles is $\mathbb{C}^2 \otimes L^2(\mathbb{R}^3) \otimes \mathbb{C}^2 \otimes L^2(\mathbb{R}^3)$. The complete wave function is $\psi = (\psi_{\alpha\beta}(\mathbf{r}_1, \mathbf{r}_2, t))$ where α and β are spinor indices, t is time and $\mathbf{r}_1$ and $\mathbf{r}_2$ are vectors in three-dimensional space.

We suppose that there are two detectors ("Alice" and "Bob") which are located in space $\mathbb{R}^3$ within the two localized regions $\mathcal{O}_A$ and $\mathcal{O}_B$ respectively, well separated from one another.

Quantum correlation describing the measurements of spins by Alice and Bob at their localized detectors is

$$G(a, \mathcal{O}_A, b, \mathcal{O}_B) = \langle\psi|\sigma \cdot aP_{\mathcal{O}_A} \otimes \sigma \cdot bP_{\mathcal{O}_B}|\psi\rangle \tag{8}$$

Here $P_{\mathcal{O}}$ is the projection operator onto the region $\mathcal{O}$.

Let us consider the case when the wave function has the form of the product of the spin function and the space function $\psi = \psi_{spin}\phi(\mathbf{r}_1, \mathbf{r}_2)$. Then one has

$$G(a, \mathcal{O}_A, b, \mathcal{O}_B) = g(\mathcal{O}_A, \mathcal{O}_B)D_{spin}(a, b) \tag{9}$$

where the function

$$g(\mathcal{O}_A, \mathcal{O}_B) = \int_{\mathcal{O}_A \times \mathcal{O}_B} |\phi(\mathbf{r}_1, \mathbf{r}_2)|^2 d\mathbf{r}_1 d\mathbf{r}_2 \tag{10}$$

describes correlation of particles in space. It is the probability to find one particle in the region $\mathcal{O}_A$ and another particle in the region $\mathcal{O}_B$.

One has

$$0 \leq g(\mathcal{O}_A, \mathcal{O}_B) \leq 1 \tag{11}$$

Remark. In relativistic quantum field theory there is no nonzero strictly localized projection operator that annihilates the vacuum. It is a consequence of the Reeh-Schlieder theorem. Therefore, apparently, the function $g(\mathcal{O}_A, \mathcal{O}_B)$ should be always strictly smaller than 1.

Now one inquires whether one can write the representation

$$g(\mathcal{O}_A, \mathcal{O}_B)D_{spin}(a, b) = \int \xi(a, \mathcal{O}_A, \lambda)\eta(b, \mathcal{O}_B, \lambda)d\rho(\lambda) \tag{12}$$

Note that if we are interested in the conditional probability of finding the projection of spin along vector a for the particle 1 in the region $\mathcal{O}_A$ and the projection of spin along the vector b for the particle 2 in the region $\mathcal{O}_B$ then we have to divide both sides of Eq. (12) by $g(\mathcal{O}_A, \mathcal{O}_B)$.

Instead of Eq (2) in Theorem 1 now we have the modified equation

$$g\cos(\alpha - \beta) = E\xi_\alpha\eta_\beta \tag{13}$$

The factor g is important. In particular one can write the following representation [8] for $0 \leq g \leq 1/2$:

$$g\cos(\alpha - \beta) = \int_0^{2\pi} \sqrt{2g}\cos(\alpha - \lambda)\sqrt{2g}\cos(\beta - \lambda)\frac{d\lambda}{2\pi} \tag{14}$$

Therefore if $0 \leq g \leq 1/2$ then there exists a solution of Eq. (13) where

$$\xi_\alpha(\lambda) = \sqrt{2g}\cos(\alpha - \lambda), \quad \eta_\beta(\lambda) = \sqrt{2g}\cos(\beta - \lambda)$$

and $|\xi_\alpha| \leq 1$, $|\eta_\beta| \leq 1$. If $g > 1/\sqrt{2}$ then it follows from Theorem 2 that there is no solution to Eq. (13). We have obtained

Theorem 3. If $g > 1/\sqrt{2}$ then there is no solution $(\Lambda, \mathcal{F}, d\rho(\lambda), \xi_\alpha, \eta_\beta)$ to Eq. (13) with the bounds $|\xi_\alpha| \leq 1$, $|\eta_\beta| \leq 1$. If $0 \leq g \leq 1/2$ then there exists a solution to Eq. (13) with the bounds $|\xi_\alpha| \leq 1$, $|\eta_\beta| \leq 1$.

Remark. A local modified equation reads

$$|\phi(r_1, r_2, t)|^2 \cos(\alpha - \beta) = E\xi(\alpha, r_1, t)\eta(\beta, r_2, t).$$

Relativistic Particles

We can not immediately apply the previous considerations to the case of relativistic particles such as photons and the Dirac particles because in these cases the wave function can not be represented as a product of the spin part and the spacetime part. Let us show that the wave function of photon can not be represented in the product form. Let $A_i(k)$ be the wave function of photon, where $i = 1, 2, 3$ and $k \in \mathbb{R}^3$. One has the gauge condition $k^i A_i(k) = 0$. If one supposes that the wave function has a product form $A_i(k) = \phi_i f(k)$ then from the gauge condition one gets $A_i(k) = 0$. Therefore the case of relativistic particles requires a separate investigation.

Noncommutative Spectral Theory and Local Realism

As a generalisation of the previous discussion we would like to suggest here a general relation between quantum theory and theory of classical stochastic processes [10] which expresses the condition of local realism. Let $\mathcal{H}$ be a Hilbert space, ρ is the density operator, $\{A_\alpha\}$ is a family of self-adjoint operators in $\mathcal{H}$. One says that the family of observables $\{A_\alpha\}$ and the state ρ satisfy to *the condition of local realism* if there exists a probability space $(\Lambda, \mathcal{F}, d\rho(\lambda))$ and a family of random variables $\{\xi_\alpha\}$ such that the range of ξ_α belongs to the spectrum of A_α and for any subset $\{A_i\}$ of mutually commutative operators one has a representation

$$Tr(\rho A_{i_1} ... A_{i_n}) = E\xi_{i_1}...\xi_{i_n}$$

If the family $\{A_\alpha\}$ would be a maximal commutative family of self-adjoint operators then the previous representation would be nothing but the usual spectral theorem [21]. In our case the family $\{A_\alpha\}$ consists from not necessary commuting operators. Hence we can call such a representation a noncommutative spectral representation. Of course one has a question for which families of operators and states the noncommutative spectral theorem is valid, i.e. when we can write the noncommutative spectral representation.

It would be helpful to study the following problem: describe the class of functions $f(t_1, ..., t_n)$ which admits the representation of the form

$$f(t_1, ..., t_n) = Ex_{t_1}...z_{t_n}$$

where $x_t, ..., z_t$ are random processes which obey the bounds $|x_t| \leq 1, ..., |z_t| \leq 1$.

From the previous discussion we know that there are such families of operators and such states which do not admit the noncommutative spectral representation and therefore they do not satisfy the condition of local realism. Indeed let us take the Hilbert space $\mathcal{H} = \mathbb{C}^2 \otimes \mathbb{C}^2$ and four operators A_1, A_2, A_3, A_4 of the form (we denote $A_3 = B_1, A_4 = B_2$)

$$A_1 = \begin{pmatrix} \sin\alpha_1 & \cos\alpha_1 \\ \cos\alpha_1 & -\sin\alpha_1 \end{pmatrix} \otimes I, \quad A_2 = \begin{pmatrix} \sin\alpha_2 & \cos\alpha_2 \\ \cos\alpha_2 & -\sin\alpha_2 \end{pmatrix} \otimes I$$

and

$$B_1 = I \otimes \begin{pmatrix} -\sin\beta_1 & -\cos\beta_1 \\ -\cos\beta_1 & \sin\beta_1 \end{pmatrix}, \quad B_2 = I \otimes \begin{pmatrix} -\sin\beta_2 & -\cos\beta_2 \\ -\cos\beta_2 & \sin\beta_2 \end{pmatrix}$$

Here operators A_i correspond to operators $\sigma \cdot a$ and operators B_i corresponds to operators $\sigma \cdot b$ where $a = (\cos\alpha, 0, \sin\alpha)$, $b = (-\cos\beta, 0, -\sin\beta)$. Operators A_i commute with operators B_j, $[A_i, B_j] = 0$, $i, j = 1, 2$ and one has

$$\langle \psi_{spin} | A_i B_j | \psi_{spin} \rangle = \cos(\alpha_i - \beta_j), \quad i, j = 1, 2$$

We know from Theorem 2 that this function can not be represented as the expected value $E\xi_i\eta_j$ of random variables with the bounds $|\xi_i| \leq 1$, $|\eta_j| \leq 1$.

However, as it was discussed in this paper, the space part of the wave function was neglected in the previous consideration. It is tempting to speculate that in physics one could have only such states and observables which satisfy the condition of local realism. Perhaps we should restrict ourself in this proposal to the consideration of only such families of observables which satisfy the condition of relativistic local causality.

Let us now apply these considerations to quantum cryptography.

4 The Quantum Key Distribution

There are quantum cryptographic protocols with one and with two particles, for a review see for example [19]. Here we shall consider the quantum key distribution with two particles. Ekert [11] showed that one can use the Einstein-Podolsky-Rosen correlations to establish a secret random key between two parties ("Alice" and "Bob"). Bell's inequalities are used to check the presence of an intermediate eavesdropper ("Eve"). There are two stages to the quantum cryptographic protocol, the first stage over a quantum channel, the second over a public channel.

The quantum channel consists of a source that emits pairs of spin one-half particles, in a singlet state. The particles fly apart towards Alice and Bob, who, after the particles have separated, perform measurements on spin components along one of three directions, given by unit vectors a and b. In the second stage Alice and Bob communicate over a public channel. They announce in public the orientation of the detectors they have chosen for particular measurements. Then they divide the measurement results into two separate groups: a first group for which they used different orientation of the detectors, and a second group for which they used the same orientation of the detectors. Now Alice and Bob can reveal publicly the results they obtained but within the first group of measurements only. This allows them, by using Bell's inequality, to establish the presence of an eavesdropper (Eve). The results of the second group of measurements can be converted into a secret key. One supposes that Eve has a detector which is located within the region $\mathcal{O}_E$ and she is described by hidden variables λ.

We will interpret Eve as a hidden variable in a realist theory and will study whether the quantum correlation can be represented in the form Eq. (12). From Theorem 3 one can see that if the following inequality

$$g(\mathcal{O}_A, \mathcal{O}_B) \leq 1/2 \tag{15}$$

is valid for regions $\mathcal{O}_A$ and $\mathcal{O}_B$ which are well separated from one another then there is no violation of the CHSH inequalities (6) and therefore Alice and Bob can not detect the presence of an eavesdropper. On the other side, if for a pair of well separated regions $\mathcal{O}_A$ and $\mathcal{O}_B$ one has

$$g(\mathcal{O}_A, \mathcal{O}_B) > 1/\sqrt{2} \tag{16}$$

then it could be a violation of the realist locality in these regions for a given state. Then, in principle, one can hope to detect an eavesdropper in these circumstances.

Note that if we set $g(\mathcal{O}_A, \mathcal{O}_B) = 1$ in (12) as it was done in the original proof of Bell's theorem, then it means we did a special preparation of the states of particles to be completely localized inside of detectors. There exist such well localized states (see however the previous Remark) but there exist also another states, with the wave functions which are not very well localized inside the detectors, and still particles in such states are also observed in detectors. The fact that a particle is observed inside the detector does not mean, of course, that its wave function is strictly localized inside the detector before the measurement. Actually one has to perform a thorough investigation of the preparation and the evolution of our entangled states in space and time if one needs to estimate the function $g(\mathcal{O}_A, \mathcal{O}_B)$.

4.1 Gaussian Wave Functions

Now let us consider the criterion of locality for Gaussian wave functions. We will show that with a reasonable accuracy there is no violation of locality in this case. Let us take the wave function ϕ of the form $\phi = \psi_1(\mathbf{r}_1)\psi_2(\mathbf{r}_2)$ where the individual wave functions have the moduli

$$|\psi_1(\mathbf{r})|^2 = (\frac{m^2}{2\pi})^{3/2}e^{-m^2\mathbf{r}^2/2}, \quad |\psi_2(\mathbf{r})|^2 = (\frac{m^2}{2\pi})^{3/2}e^{-m^2(\mathbf{r}-\mathbf{l})^2/2} \tag{17}$$

We suppose that the length of the vector $\mathbf{l}$ is much larger than $1/m$. We can make measurements of $P_{\mathcal{O}_A}$ and $P_{\mathcal{O}_B}$ for any well separated regions $\mathcal{O}_A$ and $\mathcal{O}_B$. Let us suppose a rather nonfavorite case for the criterion of locality when the wave functions of the particles are almost localized inside the regions $\mathcal{O}_A$ and $\mathcal{O}_B$ respectively. In such a case the function $g(\mathcal{O}_A, \mathcal{O}_B)$ can take values near its maximum. We suppose that the region $\mathcal{O}_A$ is given by $|r_i| < 1/m, \mathbf{r} = (r_1, r_2, r_3)$ and the region $\mathcal{O}_B$ is obtained from $\mathcal{O}_A$ by translation on $\mathbf{l}$. Hence $\psi_1(\mathbf{r}_1)$ is a Gaussian function with modules appreciably different from zero only in $\mathcal{O}_A$ and similarly $\psi_2(\mathbf{r}_2)$ is localized in the region $\mathcal{O}_B$. Then we have

$$g(\mathcal{O}_A, \mathcal{O}_B) = \left(\frac{1}{\sqrt{2\pi}} \int_{-1}^{1} e^{-x^2/2}dx \right)^6 \tag{18}$$

One can estimate (18) as

$$g(\mathcal{O}_A, \mathcal{O}_B) < \left(\frac{2}{\pi} \right)^3 \tag{19}$$

which is smaller than $1/2$. Therefore the locality criterion (15) is satisfied in this case.

Let us remind that there is a well known effect of expansion of wave packets due to the free time evolution. If ϵ is the characteristic length of the Gaussian wave packet describing a particle of mass M at time $t = 0$ then at time t the characteristic length ϵ_t will be

$$\epsilon_t = \epsilon\sqrt{1 + \frac{\hbar^2 t^2}{M^2\epsilon^4}}. \tag{20}$$

It tends to $(\hbar/M\epsilon)t$ as $t \to \infty$. Therefore the locality criterion is always satisfied for nonrelativistic particles if regions $\mathcal{O}_A$ and $\mathcal{O}_B$ are far enough from each other.

5 Conclusions

We have studied some problems in quantum information theory which requires the inclusion of spacetime variables. In particular entangled states in space and time are considered. A modification of Bell's equation which includes the spacetime variables is suggested and investigated. A general relation between quantum theory and theory of classical stochastic processes was suggested which expresses the condition of local realism in the form of a noncommutative spectral theorem. Applications of this relation to the security of quantum key distribution in quantum cryptography were considered.

There are many interesting open problems in the approach to quantum information in space and time discussed in this paper. Some of them related with the noncommutative spectral theory and theory of classical stochastic processes have been discussed above.

In quantum cryptography there are important open problems which require further investigations. In quantum cryptographic protocols with two entangled photons to detect the eavesdropper's presence by using Bell's inequality we have to estimate the function $g(\mathcal{O}_A, \mathcal{O}_B)$. To increase the detectability of the eavesdropper one has to do a thorough investigation of the process of preparation of the entangled state and then its evolution in space and time towards Alice and Bob. One has to develop a proof of the security of such a protocol.

In the previous section Eve was interpreted as an abstract hidden variable. However one can assume that more information about Eve is available. In particular one can assume that she is located somewhere in space in a region $\mathcal{O}_E$. It seems one has to study a generalization of the function $g(\mathcal{O}_A, \mathcal{O}_B)$, which depends not only on the Alice and Bob locations $\mathcal{O}_A$ and $\mathcal{O}_B$ but also depends on the Eve location $\mathcal{O}_E$, and try to find a strategy which leads to an optimal value of this function.

Acknowledgments

I am grateful to Prof. T. Hida and Prof. K. Saito for the invitation to the stimulating and fruitful conference on Quantum Information at Meijo University. This work was supported in part by RFFI 99-0100866 and by INTAS 99-00545 grants.

References

1. Quantum Information, eds. T. Hida nnd K. Saito, (World Scintific, Singapore, 1999).

2. Masanori Ohya, *A mathematical foundation of quantum information and quantum computer - on quantum mutual entropy and entanglement,* http://xxx.lanl.gov/abs/quant-ph/9808051.

3. A. Afriat and F. Selleri, *The Einstein, Podolsky, and Rosen Paradox in Atomic, Nuclear, and Particle Physics,* Plenum Press, 1999.

4. J.S. Bell, Physics, **1**, 195 (1964).

5. W.M. de Muynck, W. De Baere and H. Martens, *Found. of Physics,* **23,**1589(1994).

6. Luigi Accardi and Massimo Regoli, *Locality and Bell's inequality,* quant-ph/0007005.

7. Andrei Khrennikov, *Non-Kolmogorov probability and modified Bell's inequality,* quant-ph/0003017.

8. I. Volovich, Ya. Volovich, *Bell's Theorem and Random Variables,* http://xxx.lanl.gov/abs/quant-ph/0009058.

9. Igor V. Volovich, *Bell's Theorem and Locality in Space,* http://xxx.lanl.gov/abs/quant-ph/0012010.

10. T. Hida, *Brownian Motion,* Springer-Verlag, 1980.

11. A.K. Ekert, Phys. Rev. Lett. **67,** 661 (1991).

12. S. Wiesner, *Conjugate coding,* SIGACT News, 15:1 (1983) pp.78-88.

13. C.H. Bennett and G. Brassard, *Quantum cryptography: Public key distribution and coin tossing,* in: *Proc. of the IEEE Inst. Conf. on Comuters, Systems, and Signal Processing, Bangalore, India* (IEEE, New York,1984) p.175

14. Nicolas Gisin, Grigoire Ribordy, Wolfgang Tittel, Hugo Zbinden, *Quantum Cryptography,* http://xxx.lanl.gov/abs/quant-ph/0101098.

15. Dominic Mayers, *Unconditional security in Quantum Cryptography,* http://xxx.lanl.gov/abs/quant-ph/9802025.

16. Hoi-Kwong Lo, H. F. Chau *Unconditional Security Of Quantum Key Distribution Over Arbitrarily Long Distances,* http://xxx.lanl.gov/abs/quant-ph/9803006.

17. Peter W. Shor and John Preskill, *Simple Proof of Security of the BB84 Quantum Key Distribution Protocol,* http://xxx.lanl.gov/abs/quant-ph/quant-ph/0003004.

18. Igor V. Volovich, *An Attack to Quantum Cryptography from Space,* http://xxx.lanl.gov/abs/quant-ph/0012054.

19. I.V. Volovich and Ya.I. Volovich, *On Classical and Quantum Cryptog-*

raphy, Lectures at the Volterra-CIRM International School "Quantum Computer and Quantum Information", Trento, Italy, July 25-31, 2001 (to be published).

20. K. Fihtener and M. Ohya, *Quantum teleportation and beam splitting*, to be published in Comm.Math.Phys.

21. K. Yosida, *Functional Analysis*, Springer-Verlag, 1965.